KEEPING CHICKENS

KEEPING CHICKENS

GETTING THE BEST FROM YOUR CHICKENS

Jeremy Hobson and Celia Lewis

D&C
David and Charles

A DAVID & CHARLES BOOK
Copyright © David & Charles Limited 2007, 2010

David & Charles is an F+W Media, Inc. company
4700 East Galbraith Road
Cincinnati, OH 45236

First published in the UK in 2007
Reprinted 2007 (three times), 2008 (four times), 2009 (twice)
This revised edition published in 2010
Reprinted 2010

Text copyright © Jeremy Hobson and Celia Lewis 2007, 2010
Photography © David & Charles, except those listed on page 174
Jeremy Hobson and Celia Lewis have asserted their right to be
identified as authors of this work in accordance with the Copyright,
Designs and Patents Act, 1988.

A catalogue record for this book is available from the British Library.

ISBN-13: 978-0-7153-3625-0 paperback
ISBN-10: 0-7153-3625-8 paperback

Printed in the USA by CJK
for David & Charles
Brunel House, Newton Abbot, Devon

Publisher: Stephen Bateman
Commissioning Editor: Neil Baber
Editorial Manager and Picture Researcher: Emily Pitcher
Editor: Verity Muir
Project Editor: Caroline Taggart
Assistant Editor: James Brooks
Art Editor: Charly Bailey
Production Controller: Bev Richardson
Proofreader: Ali Smith
Indexer: Lisa Footitt

www.davidandcharles.co.uk

CONTENTS

INTRODUCTION

JUST AS A HOUSE WITHOUT
CLUTTER AND THE HUSTLE AND
BUSTLE OF PEOPLE FEELS LIKE
A MUSEUM, SO A PLOT OF LAND
WITHOUT CHICKENS CAN BE
A STERILE AND LONELY PLACE.
A GARDEN FULL OF FLOWERS
MIGHT BE BEAUTIFUL, BUT THINK
HOW SUCH A SCENE WOULD BE
TRANSFORMED BY THE GLIMPSE
THROUGH THE BLOOMS OF A
HANDSOME COCKEREL AND HIS
HAREM OF HENS.

Chickens quite literally add 'life' to your life, providing a source of joy and relaxation after a stressful day. They are relatively inexpensive to feed and keep, will not take up too much of your time and come in such a range of shapes and sizes that they will fit in virtually anywhere. Best of all, your hens will supply you with the ingredients for many varied meals – what more could you ask for?

However, there is no escaping the fact that people and poultry may, in certain circumstances, coexist uneasily, although with care and forethought most of the common problems can be avoided or overcome. Chickens and bantams can live alongside neighbours, herbaceous borders and a busy lifestyle, but it is important that the right breeds are chosen to match the situation. The correct numerical balance must also be struck, and the hen-house should be positioned so as not to cause offence. A guiding hand can often help with issues such as these but, in the absence of an experienced friend or neighbour, an easily read book can be just as useful.

To answer poultry-related questions before they even arise, *Keeping Chickens* clearly explains both the theory and the practical aspects of chicken keeping. It also includes recipes for those lovely fresh eggs, as well as providing quirky 'I never knew that' pieces of information. Who would have thought, for instance, that the Scots Dumpy was bred with short legs to stop it wandering away from the croft, or that only white eggs were seen in the UK until the arrival of Asiatic breeds in the mid-19th century? And did you know that alektorophobia is the fear of chickens?

No matter how experienced a person may be in a certain field, they cannot know everything there is to know and will still have things to learn on a daily basis. Nowhere is this more true than of keeping chickens. This book is written by two experienced poultry breeders and authors who hope both to introduce the delights of chicken keeping to the newcomer and to share their fascination in the hobby with those who are already seasoned campaigners.

RIGHT A pair of Marans strut along the top of a fence.

Understanding CHICKENS

HENS ARE THE ULTIMATE RECYCLERS. THESE OMNIVOROUS BIRDS WILL EAT ALL YOUR LEFTOVERS, GARDEN WEEDS, LAWN MOWINGS, PESTS SUCH AS SLUGS, SNAILS AND EVEN MICE, AND WILL TURN THEM INTO BEAUTIFUL FRESH EGGS WITH THE ADDITION OF JUST A SMALL AMOUNT OF CONCENTRATES AND A HANDFUL OF CORN. NOT ONLY ARE HENS USEFUL, THEY ALSO MAKE DECORATIVE, CHARACTERFUL AND CHARMING PETS. ASIDE FROM FOOD AND WATER, ALL THEY REQUIRE IS A HOUSE AND NEST BOXES, SHELTER FROM THE RAIN AND SUN, AND A DRY DUST-BATH.

A BRIEF HISTORY OF THE DOMESTICATED FOWL

The red jungle fowl (*Gallus gallus*), a native of Southeast Asia, is a tropical gallinaceous member of the pheasant family and the direct ancestor of today's domestic chicken. First raised in captivity some 5,000 years ago in India, the domestic chicken is now kept as a source of meat and eggs all over the world. Today, there are as many as 200 different breeds, some specifically developed to produce eggs, some for meat and some entirely for their beauty.

Chickens arrived in Europe in the 7th century BC; the earliest known illustration of a hen appears on Corinthian pottery dating from that time. The Romans are credited with introducing poultry to Britain when they arrived in AD 43. Chickens were associated with many superstitions by the Romans, who also used the birds for oracles. According to the Roman orator and historian Cicero, writing in the first century BC, it was considered a good omen if a hen appeared from the left, either flying or walking along the ground. The chickens' feeding behaviour was also observed when an omen was needed: the pullarius, who looked after the birds, opened up their cages and gave them a special kind of food. If the chickens ate it immediately, the omen was favourable; if they flapped about or flew away, the omen was bad.

The Roman author Columella wrote a treatise on agriculture in the 1st century AD, and gave advice on breeding and keeping chickens. He considered 200 birds the ideal-sized flock for one person to supervise. Columella suggested that white chickens should be avoided as they were easily caught by birds of prey, and that coops should lie adjacent to the kitchen as smoke from fires was beneficial to the chickens' health.

Until the mid-19th century, chickens in Britain roamed free around farms and cottage gardens, foraging for themselves and laying their eggs wherever they pleased – so they had to be hunted for. But following the arrival of Asiatic breeds such as the Cochin, with their novel brown eggs, interest in chicken keeping took off and enthusiasts started confining their hens to hen-houses with attached runs. The Victorians' passion for poultry, combined with the interests of the cock-fighting fraternity (see below), led to the development of new breeds and, eventually, to the establishment of poultry clubs and exhibitions.

COCK-FIGHTING

Interest in cock-fighting was partly responsible for the encouragement of poultry breeding in Britain, in particular the development of the Old English Game fowl. The sport originated in China and was particularly cruel – the cockerels were specially bred to be aggressive and had artificial razor-sharp points known as cockspurs fitted to their legs. The birds pecked and maimed each other with their spurs until one died, although the victor often subsequently suffered the same fate as a result of injuries. The fights were popular with spectators who wagered large sums on the outcome. The sport was made illegal in the UK in 1849, but is still practised in Asia and in some European countries, as well as in the state of New Mexico (though it is banned elsewhere in the USA).

RIGHT Looking after the chickens was traditionally the responsibility of the farmer's wife, but children could help too.

Chickens in Religion

Chickens feature in many religions. In India, the cockerel is an emblem of Karthikeya, son of Lord Shiva, the Hindu god of destruction, while Hindus in Indonesia use chickens in their cremation rites. Throughout the ceremony a chicken is tethered by its leg to ensure that any evil spirits enter it rather than any family members present. After the ceremony, the chicken is taken home to resume its normal life.

The ancient Greeks believed that cockerels were so brave even lions were afraid of them, and they are found as attributes of the gods Ares, Heracles and Athena. In the ancient Persian cult of Mithras, the cock bird was a symbol of the divine light and a guardian against evil.

In the Bible, Jesus prophesied his betrayal by Peter: 'And he said, I tell thee, Peter, the cock shall not crow this day, before that thou shalt thrice deny that thou knowest me' (Luke 22:34). As a result, the cock became a symbol of both vigilance and betrayal in Christian societies.

single comb

rose comb

v-shaped comb

walnut comb

pea comb

buttercup comb

COMBS, WATTLES AND EAR LOBES

One of the most distinctive features of a chicken is its comb, which, with its wattles (fleshy appendages under the beak), functions as its cooling system. Birds cannot sweat, so the chicken cools itself by circulating blood through its comb and wattles, from which body heat radiates.

There are several types of comb, the most common of which is the single. This has a number of serrations and stands upright on the head, although in some breeds with very large single combs the back part tends to droop to one side. In contrast, the rose comb is low and broad, lacks points and tapers off into a 'spike' at the back. There are also v-shaped combs, walnut or cushion combs, pea combs and buttercup combs.

Surprisingly, ear-lobe colour governs egg colour. Generally, if the hen has white ear lobes she will lay white-shelled eggs, whereas red ear lobes indicate brown-shelled eggs. One exception to this rule is the Araucana, which has red ear lobes but lays blue- or green-shelled eggs.

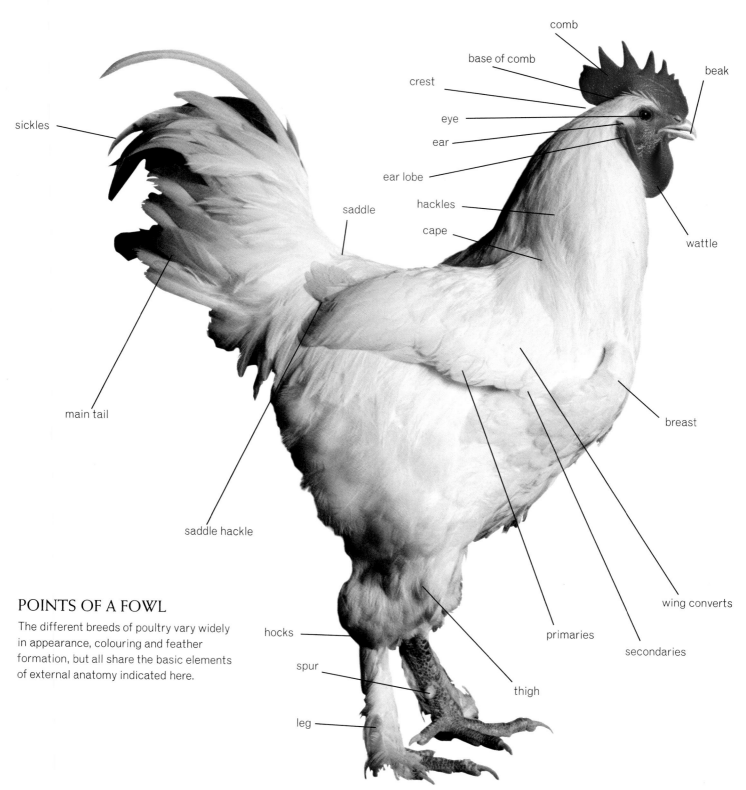

comb

base of comb

crest

beak

eye

ear

sickles

ear lobe

wattle

saddle

hackles

cape

main tail

breast

saddle hackle

wing converts

POINTS OF A FOWL

The different breeds of poultry vary widely in appearance, colouring and feather formation, but all share the basic elements of external anatomy indicated here.

hocks

primaries

secondaries

spur

thigh

leg

barred

laced

spangled

pencilled

FEATHERS

Chicken feathers come in many colours and patterns, and provide both insulation and waterproofing. The condition of a bird's feathers is maintained through preening, which takes up a large proportion of its day and involves dust-bathing and applying oil from a gland at the base of the tail. From a zoological point of view, feathers are composed of keratin, like hair and horn, and correspond to the scales of a fish, as they similarly overlap one another.

Poultry breeds are categorized as hard- or soft-feathered. The former are those originally bred for cock-fighting, which have smooth, close-fitting plumage. Not all birds have the same amount of feathering – some breeds have feathers right down their legs and may also have feather beards and crests.

The hackle and saddle feathers of cock birds can be distinguished from those of hens in that they tend to be pointed and shiny. The crest feathers of cock birds are the same shape as its hackles, while in hens they are shorter and broader. Cocks also have sickles, which are the long, curved feathers on either side of the tail.

Distinctive feather patterns can be used to identify the many varieties of chicken:

○ **Barred** feathers have alternate transverse markings in two distinct colours.
○ **Frizzled** feathers curl and curve outwards and forwards in a random manner (in the UK the Frizzle is a recognized breed – see page 60).
○ **Laced** feathers have a border of a contrasting colour all around the edge.
○ **Mottled** feathers have a variable percentage of black feathers tipped with white.
○ **Spangled** feathers have a distinct marking of a contrasting colour, usually in a v shape at their tip.
○ **Pencilled** feathers include several types of lines or markings, although most commonly they have narrow concentric linear markings following the contours of the edges.

THE PECKING ORDER

Chickens are gregarious birds that live together as a flock. The harmony of the group is maintained by the observation of a hierarchy, termed the pecking order, which governs where each chicken stands within the flock.

The pecking order starts with the 'top bird' and extends down to the youngest and weakest, which survive as best they can. The top bird is usually an old hen, although sometimes it is the most aggressive bird, and the rest of the flock will defer to her, often giving way at the food container.

When new birds are introduced to an established flock there are always problems because of the upset to the pecking order. A hen that spots a newcomer will utter a single warning croak, alerting the rest of the flock. If the birds were living in the wild the newcomers would be seen off, but because this is not possible in a run they may be bullied unmercifully until they are eventually accepted. One of the points in favour of keeping a cockerel is that he will stop the older members of the flock bullying the junior hens. To help protect new birds from becoming henpecked in this way, introduce them at night and never one at a time.

Without an established flock to show them what to do, new birds can be remarkably stupid about knowing that they should go into their house at night and roost on the perches provided, often preferring to fly up into trees or to teeter on the hen-house roof. When you buy new stock, it is a good idea to make sure that all the birds are perching happily for the first few nights; if they are not, lift them on to their perches until they learn to do this themselves.

INTERESTING FACTS

Chickens are probably the most widespread of all domestic animals. There are approximately 29 million chickens in Britain, 271 million in the European Union and 400 million in the USA.

A chicken has a body temperature of 39–39.5°C (102–103°F).

A chicken's heart beats at the rate of 280–315 times a minute.

The average lifespan of a chicken is 5–7 years, although 20 years is not unknown.

In 1925, the average chicken laid 100 eggs a year; in 1979, a White Leghorn set the world record by laying 371 eggs in 365 days.

Alektorophobia is the fear of chickens.

Incubation starts when the egg reaches a temperature of 30°C (86°F).

The largest chicken egg ever recorded was laid by a Black Minorca in 1896. It weighed 340g (12oz), had a circumference of 23cm (9in) and contained five yolks.

The heaviest chicken on record weighed 10.5kg (23lb 3oz).

What to CONSIDER

WHEN DECIDING ON SPECIES AND SYSTEMS OF MANAGEMENT, YOU FIRST NEED TO BE CLEAR ABOUT YOUR AIMS. ASK YOURSELF WHETHER YOU ARE INTERESTED IN KEEPING CHICKENS FOR PROFIT, SELF-SUFFICIENCY, OR PURELY FOR PLEASURE. YOU ALSO NEED TO ASCERTAIN HOW MUCH TIME AND SPACE YOU HAVE AT YOUR DISPOSAL. LITTLE OF EITHER WILL NOT NECESSARILY PRECLUDE YOU FROM KEEPING A SMALL PEN OF BANTAMS, BUT IT WILL CERTAINLY REMOVE LARGE, FEATHER-LEGGED AND POTENTIALLY MORE DEMANDING VARIETIES FROM THE EQUATION.

SPACE

The theory behind keeping free-range poultry is wonderful but the practice is not always quite so simple. While many people's ideal is to have chickens scratching at the back door, picking at scraps and living a contented life, in reality things can be very different.

Although free-ranging poultry will eradicate insect pests and slugs in the garden, some of their habits are less welcome. Chickens and bantams love nothing more than a good dust-bath to help rid themselves of parasites, and consider a well-prepared seedbed ideal for the purpose. To their mind, bark mulch that has been carefully placed around plants is scratching heaven. For these reasons alone, you may prefer to keep your chickens confined to a run. A run will also prevent your hens from laying in the shrubbery (see page 21) and, if well constructed, will protect them from neighbourhood dogs and any predators.

The amount of space needed in the run depends on whether your chickens can be given some free-range, but generally a run should be as big as is practicable. In the interests of hygiene, two runs can be advantageous – one in use and the other 'resting' – although this does double the area needed. For more on the amount of space needed for a house and run, see Housing & Cleaning, page 98.

LEFT A good chicken run will keep your birds safe while also preventing them from ruining your garden. ABOVE RIGHT The behaviour of dogs around chickens varies according to temperment.

CHILDREN AND PETS

While chickens make excellent pets for children, some breeds are known to be more aggressive or flighty than others. Therefore, it may be sensible, especially if you have toddlers, to avoid keeping cock birds of these breeds. Instead, opt for bantams, whose smaller size is also more likely to appeal to youngsters.

If you already have other pets, you should consider how they are likely to react to the arrival of a flock of chickens in their territory. Cats generally get on very well with hens, and most dogs will accept the birds as part of the family and show no interest in them. However, some dogs may attack and kill them – this is especially true with some types of bantam, which seem to have a more gamey smell. Rabbits and guinea pigs will live happily near hens but should not be put in with a flock that is fenced in a small run. Hens can be bullies and tend to peck defenceless creatures either from curiosity or boredom. Hens and ponies are an excellent mix – ponies will enjoy the company and the hens will scratch around, eating up all the dropped food and hay seed.

YOUNG CHICKEN KEEPERS

Keeping chickens is an excellent hobby for children, teaching them responsibility through feeding and general care, and also about food production and biology. As hens lay eggs, they can also learn about budgeting and accounts, as well as how to cook the end product.

To further childrens' interest in keeping birds, the Poultry Club of Great Britain organizes the Junior Certificate of Proficiency in Poultry Husbandry. This examination can be taken by children aged under 16, is free and open to both members and non-members of the club. The syllabus is designed to establish a basic knowledge of stockmanship and poultry husbandry.

The Poultry Club also runs the annual Junior Fancier of the Year Award at the National Championship Show, but to qualify for this a youngster must first have won a Champion Juvenile award at one of that year's major regional shows. The Junior Fancier of the Year is judged with special regard to the youngster's knowledge and handling competence, as well as the quality of his or her exhibit.

LEFT AND ABOVE Looking after chickens can teach children a lot about responsibility.

CHICKENS & THE GARDEN

If you want to keep free-range chickens but at the same time would like to maintain a reasonably tidy garden, there are several tricks you can employ:

Choose bantams with feathered legs, such as Pekins, as these won't scratch up the grass on your lawn as much as other breeds.

Lay chicken wire flat on your herbaceous or annual beds in winter. The plants will grow through the wire but the hens won't be able to scratch them up.

If dust-bathing in your beds becomes a problem, try tempting your birds away from them with an irresistible box of dry sand or peat. If you place it in a sunny position they may prefer it to the flowerbeds.

If your hens lay outside rather than inside the cosy nest-boxes you have provided, don't let them out of their run until mid-morning, by which time most should have finished laying. Some birds may still wait to visit their favourite spot, in which case you will have to spy on them, as they won't go near it if they know they are being watched.

Fertilize your plants with the manure your chickens produce — around 100g (3½oz) per bird per day. It is high in nitrogen, which keeps foliage green and makes an excellent lawn feed when diluted in water. However, as it is high in ammonia it will burn plants if applied direct to beds. Instead, add it to your compost to allow it to break down first — it will also help to heat the heap and aid rotting.

CHICKEN BEHAVIOUR

Chickens do the silliest things – or at least that's how it sometimes seems. In fact, there is usually method in their madness and that sudden eccentric dash across the lawn where, seconds earlier, they had been contentedly clucking and scratching, was probably the result of a discovery of newly hatched insects or grubs, and not a crazy brainstorm. Pecking birds that abruptly stop and look upwards are not, in the way of 'Chicken Licken' of children's story-book fame, looking to see whether the sky is falling in, but have in all probability just noticed the silhouette of a bird of prey passing over, or at least a raptor look-alike such as a crow. Despite centuries of domesticity and evolution, their primeval awareness of potential danger has not entirely disappeared and is evident in their daily life.

There is a natural and often seasonal order to chicken behaviour. As with most societies, each poultry flock creates a hierarchy and, while the male is generally accepted as leader of the group, if there is no cockerel one of the more dominant hens will take his role. When discovering a dainty titbit, she may or may not choose to offer this to the others as a cockerel would and will keep them all in order by means of a judicious peck. Not for nothing has the expression 'pecking order' become an accepted part of the English language (see page 15).

Also originating from the chicken world is the word 'broody'. In human parlance it can mean 'sullenly thoughtful', but more often describes the attitude of 'a woman wanting pregnancy'. The instinct of most hens, especially those of the heavier types described in the Choosing the Breed chapter beginning on page 32, is to lay a clutch of eggs and then brood them to ensure the continuation of the species. As Samuel Butler wrote in his 1878 classic *Life and Habit*, 'It has, I believe, been often remarked that a hen is only an egg's way of making another egg'. Left to herself, a hen would lay perhaps 13–15 eggs (almost always an odd number because they fit more comfortably into the shape of a nest and under the hen's 'brood spot' than an even-numbered clutch) and it is only because her eggs are continually removed by her keeper that she fails to go broody more quickly. For more about broody hens, see pages 140–141.

Most hens stop laying when, at the end of summer and beginning of autumn, they start to moult (see page 160). It can be alarming to see your chicken run slowly filling with feathers as the old ones fall out and are replaced by new, but this is merely nature's way of ensuring that the bird has body feathers that will protect it from wind and rain, and wing feathers that could enable the chicken to fly away from predators if the need arose.

RIGHT Without a cockerel a flock of hens will establish their own 'pecking order'.

PREPARING YOUR GARDEN

Although it is impossible to say how many birds an average-sized garden will hold, even the smallest plot of grass should be sufficient for a combined house and run, which must be moved on a regular basis in order to prevent damage to the turf and a possible build-up of parasites. No matter how large the area available, you should be careful never to overstock it. Recent research has shown that high stocking densities interfere with a chicken's ability to function normally and that this cannot be mitigated by other factors such as good husbandry and management.

With care, you can keep a flock of backyard chickens without any ill effect to themselves or the garden. However, you cannot simply open the door of your chicken run and allow its inhabitants free-range without first checking at least one or two essential criteria, perhaps the foremost of which is how much you love your garden? Whatever the answer, nothing should stop you from including the colourful living sculptures that are chickens. A dust-bath might help keep them off the flowerbeds. In its simplest form, it is nothing more than a dusty hole created by the birds themselves, either between the exposed roots of a tree or under the garden shed. An artificial equivalent can be made by placing a mixture of dry earth, fire-ash and sand within the confines of a wooden frame, or even an old car tyre painted to match its surroundings.

Check for any holes or gaps in the boundary fences: you might enjoy the sight of busy free-ranging chickens in your patch,

but your neighbours might not share your opinion. Perhaps more important than your birds escaping through such apertures, is the danger of neighbouring dogs and predators gaining access. For this reason, unless your garden is well fenced, it is probably wise to let your birds have free-range only when you are around to keep an eye on them.

ABOVE RIGHT Unless you intend to breed from it, you won't need to keep a cock bird with your flock.
RIGHT Always check to make sure that your chickens can't escape into your neighbours' gardens.

ABOVE Chickens have distinct personalities and watching their activities while they roam around outside the hen-house will help you get to know them better.

DAILY ROUTINE

Although chickens and bantams are not time-consuming to look after, they do appreciate a routine and like to be let out of their house and fed at regular times. Keeping to a daily pattern is important, so before you decide to make any purchases it is vital that you consider how you will carry out the apparently simple tasks of shutting the chickens up at night and letting them out in the morning.

In the winter months many people leave for work in the dark before the birds come down off the roost. Will there be anyone at home to let your chickens out once daybreak comes? If so, can you rely on them not to forget this simple chore? In the middle of summer there may be problems at the other end of the day, as your chickens may still be scratching around when you are ready for bed. If the birds are kept in a secure pen this may not be a problem, but if they are free-range there is a strong possibility that they will be taken by a nocturnal predator.

Ideally, chickens should be fed little and often. However, in a busy household this is not always practical, so feeding morning and evening is the best option. Don't be tempted to leave too much food: not only is this wasteful, but pellets or mash can go sour if left in damp conditions and the practice will certainly encourage vermin. If food and water are given ad lib in the hen-house, the floor covering will get scratched into the feed and water will be spilt, increasing the likelihood of fungal disease.

Other parts of the daily routine include scraping the previous night's faeces from the droppings board, checking that no eggs have been broken in the nest-boxes and changing damp litter around indoor drinking vessels. In the winter, drinkers need to be kept free of ice. For more information on cleaning and maintenance routines, see Housing & Cleaning, page 98.

Although none of these chores takes long, there is no point in keeping chickens if you cannot spend some time watching and enjoying their antics. In addition to providing relaxation at the end of a busy day, time spent leaning on the gate is essential in order to keep a check on your birds' well-being.

WHEN YOU GO AWAY

When you are starting your adventure of keeping chickens, bear in mind that you will need other people's assistance when you go away. If you live among like-minded neighbours, it is not too onerous a task for one of them to pop round each morning and evening to feed and water your birds. However, if you live down a lonely track out in the country, finding someone to take on the job may not be so simple.

Equally, a small flock of birds is not likely to cause much bother for anyone taking over their care, but if you have several different breeding pens, incubators and rearing cages it is probably too great a responsibility for a neighbour or family friend. If you only have one annual holiday, you can arrange this at the least busy time in your poultry year, but if you go away regularly the size and demands of your flock should reflect this.

It is also important to plan the pen and poultry shed layout with others in mind. A gate that falls off its hinges and traps your ankle every time you go through it might not bother you, but you can be sure that outside helpers will be less enthusiastic the next time you call on their assistance. You could also consider setting up an automatic watering system. There are several of these on the market that are inexpensive and easy to install, and they will save both you and your stand-ins some time. Commercially produced automatic pop-holes are also available, although if you are creative it is easy enough to design and build one yourself that is raised and lowered by the use of a small motor and timer. If you do install such a device you will still need someone to check that the system hasn't failed and that the birds haven't been locked in all day or out at night.

BELOW If you go away make sure someone will be available to look after your chickens.

COST

Aside from the birds themselves, the initial costs of keeping chickens are those of the capital equipment – housing, wire netting, fence posts, feeders and drinkers. If you intend breeding more than the odd clutch of eggs (which can normally be done very successfully when one of the hens becomes broody), you might also consider the acquisition of a small incubator and artificial brooder.

Look after it well and there is no reason why such equipment should not remain sound and serviceable for a good 20 years or more, so the initial outlay is minimal when broken down over such a time. When it comes to choosing equipment, the less expensive alternatives may not last long, but it is worth weighing up all the options. For example, galvanized-metal drinkers and feeders are expensive but will outlive the less costly plastic ones; on the other hand, plastic is easier to keep clean.

Prices of stock will vary depending on whether you choose commercial hybrids or pure breeds, the latter usually being more expensive. Point-of-lay stock (see Where & When to Get Your Chickens, page 90) will also cost more than youngsters that need bringing on, although you do need to take into account the extra time and food required to raise young birds to maturity. If you simply want an inexpensive source of eggs, see if you can buy layers from an intensive farm. These birds are kept by the farms for their first laying season only and so have many more years of egg-laying ahead of them.

Provided that you maintain a reasonable common-sense attitude towards the question of hygiene, it is unlikely that your flock will encounter any serious disease that might require the services of a veterinary surgeon. That said, is is always possible that some expensive vet's bills could be encountered at some point.

The daily running costs of keeping chickens are low, and the purchase of pelleted feeds, cereals, grit, vitamins and floor litter will generally amount to very little. You may want to have electric lighting in the hen-house for the winter months or power points from which to run an incubator, so remember to factor in the initial expense of installing these as well as the additional running costs on your electricity bill.

Some of the costs of keeping chickens can be offset by selling eggs, chicks and surplus birds, and, if you are artistic, even chicken-inspired craftwork. However, in reality, unless you keep sufficient numbers of birds to supply your customers with eggs all year round, chicken keeping can only ever be a hobby rather than a viable business venture. Never lose sight of the fact that it is almost always less expensive to buy eggs from your local supermarket. However, for most people the excitement and appeal of collecting their own new-laid eggs from a sweet-smelling, hay-filled nest-box more than makes up for this.

BY-LAWS AND NEIGHBOURS

Before you embark on keeping chickens, check whether any local authority or other regulations are in force that prohibit the practice in your area. If this is the case, either accept that your project is doomed to failure or consider moving house!

It is also important to explain your plans to any immediate neighbours to keep on the right side of them. If you do not, you might find yourself involved in a neighbourhood quarrel or even a court case that no amount of eggs offered as a sweetener will rectify.

From an environmental health point of view, rats and other vermin must be kept under control by the use of traps or poison.

It is far better to take such preventative steps from the outset rather than wait until any complaints are officially registered. If this happens, you may be legally forced to get rid of your stock, and if you fail to do so are quite likely to be prosecuted. Likewise, the owners of a loudly crowing cockerel could find themselves subject to a noise-abatement order. Hens will lay just as well in the absence of a male. In fact, in some instances a cock bird can actually be a deterrent to laying (see page 30).

Depending on the size of your intended set-up, it may be necessary to obtain local authority planning permission if you want to construct a combined chicken shed, veranda and food store, or even showing/isolation pens. In most cases it all comes down to overall size, whether or not the new building will be attached to an existing one, and its proximity to any neighbour's boundary. Finally, any fencing around your property must be good enough to prevent your birds wandering off your land. In most countries there is a legal requirement for you to protect other people's property from your livestock.

RIGHT A busy flock of hens can bring any garden to life.

KEEPING COCK BIRDS

Technically, a cockerel is a male bird before it reaches its first adult moult (this usually takes place at around 18 months), after which it becomes known as a cock. However, generally the term cockerel denotes a young male under 12 months. Traditionally, the cock bird holds a time-honoured place in a small flock, protecting his hens, announcing the new day and helping to produce the next generation. But apart from the fact that a cock bird is necessary for breeding purposes, and that he looks wonderfully colourful as he struts his stuff, is there any real reason to keep one?

LEFT A fine example of a single-combed cockerel.

CHOOSING A PET COCK BIRD

The cockerels of some breeds are more aggressive than others, so it may be wise to avoid these if you have young children. Generally, the males of the heavier breeds, such as Maran, New Hampshire Red, Orpington, Plymouth Rock, Rhode Island Red and Sussex, are calm, placid and friendly. While the extremely heavy breeds, such as Brahma and Cochin, also make excellent pets, they may be too big for children who want to carry them around. In this case, it may be worth considering bantam varieties if these exist.

Although they were originally bred for fighting, most of the game cocks are extremely placid and friendly towards people, although less so towards other birds and animals. Faverolles, Silkies and Wyandottes are also all known to be non-aggressive towards people, including children.

LEFT A handsome cock bird keeps watch over his hens. RIGHT A Cream Legbar cockerel: this auto-sexing breed produces eggs with blue-green shells.

WHY HAVE A COCK BIRD?

Many people believe that hens will not lay without a cock bird or that they need one to watch over them, but this is untrue. In the absence of a cock, hens will find their own pecking order, or hierarchy, with one of the more dominant birds taking on the male's role. However, there is no doubt that a cock bird enhances the pleasure of keeping chickens, even where there is no real practical reason to have one. There is no scene more pleasant than an elegant-looking male busily scratching around and drawing his hens' attention to an interesting titbit, then standing proudly over them while they peck excitedly at the treasures he has found.

While a cock bird may look magnificent, he will lay no eggs, will disturb your neighbours with his early morning crowing and will eat a lot of food. More seriously, egg production can go down if the hens are stressed and injured by a cock bird's attentions during mating. In particular, his spurs and claws can damage their backs and he will also pull out their head feathers while trying to hold on. The more submissive the hen, the more likely it is that the male bird will try to mate with her, and so the more she will suffer from broken feathers and scratches as a consequence of his attentions.

OVERCOMING PROBLEMS

Some breeds of chicken are known to be more aggressive than others, so if you decide to keep any of these it will pay not to include a cock bird in the flock unless absolutely necessary. Certain breeds, such as the Ko-Shamo, also crow less loudly than others, something that might further influence your choice. Although it is impossible to stop a cock from crowing altogether, some preventative measures will work with varying degrees of success.

'CURING' CROWING

Most complaints against cock birds are because they crow at dawn, which can be particularly annoying in the summer months when it gets light very early. A cock needs to raise his head before he can crow, something he will be less inclined to do if the hen-house is a small coop and the perch is fixed high. However, this method of prevention works only when he is actually on the perch and so will simply delay crowing until he is let out rather than curing it altogether.

It is sometimes possible to prevent a cock from crowing too early by creating a shutter system, whereby the chicken-house is made totally dark. If you do try this method, make sure there is still adequate ventilation and fresh air for all the roosting birds. Another option, but one that is time-consuming, is to remove the cockerel each evening and house him in a box or specially constructed cage at the rear of a garden shed, garage or similar outbuilding, well away from your house and those of your neighbours.

PREVENTING DAMAGE TO THE HENS

The simplest way to prevent a cock injuring hens is to ensure that the flock is large enough for the cock to divide his attentions between all the hens rather than concentrating on just a few. In a breeding pen where only one cock and three or four hens of one variety are required, introduce a few extra hens, preferably ones that lay a different-coloured egg so that the correct eggs can be identified for hatching. A ratio of one cock to ten hens seems to alleviate the problem in most breeds, but it is interesting to note that sometimes one or two of the hens are not fertile because they are not compatible with that particular cock bird. Another option is to purchase lightweight canvas saddles, which are attached to the hens' backs to protect them from damage during mating.

It is also possible to trim the spurs of a cock bird, either with a strong pair of clippers normally used for trimming dogs' claws, or with a hot blade that cauterizes as it cuts (bleeding is always a possibility when an inexperienced person attempts the former method). A traditional method of removing spurs uses a potato that has been baked in the oven for an hour and then cut in half. Protect the cockerel's legs and feet by wrapping them in a towel or rolled sheets of kitchen paper, then push each half of the potato on to a spur. Hold the potato halves in place for five minutes with the aid of a thick gardening glove, then remove them. The spurs should then come away with a firm but gentle push and a quick twist.

SELECTING A BREEDING COCK

The cock bird is the most important member of the breeding pen, and if his background is doubtful many of his faults will be transferred to his offspring. Therefore, if he is to have any hope of producing good chicks, he must himself have come from quality breeding stock. He should be of the type outlined in the breed standards, well muscled and firm to handle. His eye colour should be bold and even, and should surround a perfectly black pupil. The comb should also conform to the breed requirements and his legs should be clean and strong – without sound legs and feet, he will find it difficult to mount and mate with his hens. Age is also important. While a cockerel is sexually mature from around six months, it is better to choose a bird that was bred in the previous spring or early summer.

DOUBLE MATING

The method of breeding known as double mating requires two distinct strains of birds: one to produce exhibition males (this strain is called cock breeders) and the other for exhibition females (pullet breeders). It is usually carried out to create a particular plumage type or colour, good examples being the Partridge Wyandotte and Pencilled Hamburg.

In the Partridge Wyandotte, the breed standards require males to have a solid black breast, something that is achieved by crossing indistinctly marked hens with exhibition-standard cock birds. The female chicks produced in this line are not suitable for showing. For their part, exhibition pullets require pencilling, a marking that is produced by crossing cock birds that display a mixture of brown and black in their breast plumage with exhibition-standard hens. Like the females produced from the cock-breeding strain, the male offspring of pullet breeders are not suitable for exhibition.

If you decide to keep a breed that requires two strains, your selection of cock birds must reflect this (see Breeding Chickens, page 136). In addition, you will need at least two runs so that the separate lines can be kept apart.

RIGHT It may not always be practical to keep a cock bird but their elegance and charm can be highly rewarding.

Choosing the
BREED

CHICKENS ARE SUCH CHARACTERS AND SO
ATTRACTIVE THAT IT IS EASY TO FALL IN LOVE WITH
EACH AND EVERY BREED. HOWEVER, BEFORE MAKING
A DECISION ON WHAT TYPE BEST SUITS YOU, IT IS AS
WELL TO SPEND TIME AT SHOWS AND EXHIBITIONS,
TAKING NOTES AND TALKING TO BREEDERS. PART OF
THE PLEASURE OF KEEPING CHICKENS IS KNOWING
ABOUT THE VARIOUS BREEDS AND HOW TO
DISTINGUISH BETWEEN THEM.

PURE BREEDS AND HYBRIDS

There are countless pure breeds of large fowl to choose from. Many of the old breeds have excellent qualities both as meat-producers and egg-layers, especially when they are kept in non-intensive management systems. Such birds may cost more initially, but they bring the pleasure of knowing that you are ensuring the future welfare of a recognized breed. As the chicken-keeping bug takes over your life – and it surely will – you might also consider showing your stock, an option that would not be possible if you opted for hybrids.

A hybrid is the result of a cross between two or more breeds. Such a bird will not breed true and reproduce chicks in its own likeness. Instead, for more of the same type you will have to go back to the original parent stock and breed from them again. Of course, you could keep an ever-changing flock of hybrid birds by periodically introducing unrelated cock birds and breeding them with your nucleus of hens. However, if you do this, the desirable points so obvious in the original stock are likely to be be diluted in the subsequent generations. On the plus side, hybrid birds tend to be more vigorous and productive than pure breeds, which is why the vast majority of modern commercial birds are of hybrid stock. For more details of hybrids and pure breeds, see pages 36–39.

BANTAMS

The small breeds known as bantams are the obvious choice where space is limited. It is possible to find most of the large breeds of poultry in a bantamized version, although some birds in this category are 'true' bantams, which means they do not have a larger counterpart. True bantams are kept mainly for ornamental or exhibition purposes, or as pets, and will not provide you with a year-round supply of eggs.

While some of the true bantams are compact, clean-legged birds that are easy to keep, others may need special care owing to their beards, crests, long sweeping tails or heavily-feathered legs and feet. The same problems could arise when choosing bantamized types of large fowl, but as there are many more easily manageable breeds to choose from in this category, it should be possible to find one that suits your individual circumstances perfectly. For more about bantams, see pages 40–1.

MEAT, EGGS OR EXHIBITION?

Many of the heavy breeds of poultry were originally produced to fulfil two purposes: surplus cockerels were fattened for the table, while the pullets were kept for their egg-laying abilities (and at the end of their productive years were also killed and eaten). Most of these dual-purpose breeds remain a good choice for the beginner, as they make excellent pets and become quite tame with careful daily handling. They are usually good-looking birds and are also less flighty than many of the egg-laying breeds, so require less in the way of high fencing. Any of the pure breeds can also be exhibited in poultry shows.

TABLE BREEDS

Modern-day broilers are usually hybrids based on the White Rock and Cornish Game, and score over the pure breeds for their ratio of meat to bone, their speed of growth and their ultimate weight. Despite this, it is generally thought that the meat of these hybrids is bland and tasteless in comparison to more traditional pure types of table bird such as the Crève-coeur and Indian Game.

As they are large and heavy in stature, any of the table breeds can be kept with the minimum amount of fencing. In addition, their placid nature makes them ideal pets and companions, and means that they are gentle towards each other. Despite being developed for meat, some breeds also lay surprisingly well and have a natural tendency towards going broody, which is a good thing should you consider breeding from your own stock. However, one point to bear in mind is that they tend to become overweight if they are not allowed free-range and/or are fed a diet too high in calories.

LAYERS

Most commercial hybrid layers originate from the Rhode Island Red and White Leghorn, as both breeds are known to produce a constant supply of eggs. To produce an auto-sexing breed (one in which the sex of a day-old chick is easily discernible from its markings), the Legbar was developed by crossing a Brown Leghorn with a Barred Rock. The sex of day-old Legbar chicks is easily seen, as the females have clear barring on the head and back, while the cockerels have poorly defined barring. The cross between a Rhode Island Red cockerel and a Light Sussex hen also produces auto-sexed chicks – day-old pullets are brown and cockerels are silver. These hybrid strains were popular in the 1940s, and although they are now much in decline, they would be a wonderful challenge for a novice should an egg-layer be all that is required.

Pure layer breeds are often classified as 'light', and as such they tend towards flightiness and never become quite as tame as other breeds. This is important to consider when there are children around or if you want a pet bird. It may also be necessary to build taller sides to their run or to top it with nylon netting. Another alternative would be to clip one wing of each bird, although if you do you will not be able to exhibit them until after their next moult.

HOW MANY?

When deciding on the number of birds to purchase, you first need to be clear about your expectations. Do you want a regular turn-around of table birds or a year-round supply of eggs? Do you want to rear chicks? Is it your intention to produce exhibition birds?

Generally speaking, six hens of a breed known for its laying capabilities will provide sufficient eggs for a family of four. At certain times of the year, such as late summer/ early autumn when the birds are in moult, egg production will drop, but it shouldn't cease altogether. Hens tend to lay more sporadically in the winter months because of the reduced daylight hours, but if they are kept in draught-proof surroundings in a sunny location and are adequately fed, they should be less affected by the shorter days. If egg production does tail off, set up a light bulb in the hen-house and switch it on for a while at dawn or dusk to extend the 'daylight' hours (see page 116).

ABOVE As attractive as heavily feathered birds may be, their added adornments may cause day-to-day management problems.

HYBRIDS

UNTIL RECENTLY, EGG-LAYING AND MEAT-PRODUCING HYBRIDS WERE GENERALLY BRED WITH MONEY IN MIND AND SCANT REGARD TO THE WELL-BEING OF THE BIRDS. FORTUNATELY, THE SITUATION IS CHANGING AND, DUE TO THE UPSURGE IN AMATEUR POULTRY KEEPING, HYBRIDS ARE HAVING A BETTER LIFE. AS FAR AS THE CASUAL KEEPER IS CONCERNED, HYBRIDS ARE LESS EXPENSIVE TO BUY THAN PURE BREEDS AND WILL QUICKLY LAY EGGS OR PRODUCE MEAT.

Hybrid hens can lay up to 280 eggs a year, and the vast majority of modern commercial birds are hybrid stock, though many have names more reminiscent of soft porn actresses (Amber Star) or Wild West movies (Gold Star Ranger) than the poultry world! Hybrid laying poultry have been produced in different ways but have always been based on parental strains selected to complement one another genetically. This procedure is known by some commercial producers as 'nicking'; a practice that tests different strains so that the best could be used for producing the commercially marketable end product.

Because producers provide many variations on a theme, similar-looking hybrid layers may be offered for sale under a multitude of names. With feather colouring including gold, black, silver, white, speckled and blue, it would be possible to 'mix-and-match' and colour co-ordinate your poultry flock with the flowers in your garden!

The majority of hybrids have been developed from the traditional poultry breeds. A cross between a Rhode Island Red and a Light Sussex is a common one, as is that of a Rhode Island and a Barred (Plymouth) Rock. These matings are also sex-linked and, depending on the parentage, it should be possible to identify the sex of the chicks at a day old. Mating a Rhode Island Red cockerel with a Light Sussex hen will produce females of a golden-buff colour and males of a silver-white. Adult plumage will also vary depending on the way a mating is made. Black Star hybrids are created using a Rhode Island Red cockerel with a Barred (Plymouth) Rock hen, whereas putting a Barred Rock male with a Rhode Island female will produce what is commercially known as the Speckled Star.

Best known of all egg-laying hybrids in the UK is the Black Rock; predominantly black, but with chestnut colouring around the neck. It can be bought from many agents

ABOVE AND LEFT The colour of an egg yolk is directly associated with the hen's diet: a regular, natural supply of greenstuffs will almost always produce a deeper yellow yolk.

throughout the UK but is bred only by Peter Siddons at Muirfield Hatcheries in Scotland. It is ideal for the more inhospitable climates where other commercial breeds may not fare so well. In the US, the nearest equivalent is the Bovans Nera, but the company Aviagen produces three other well-known hybrid strains: Arbor Acres, Ross and L.I.R.

Traditional hybrids bred for the table were created by use of the White Rock and Cornish Game, the genes of which still feature prominently in today's meat-producing birds as they score over the pure breeds for their ratio of meat to bone, speed of growth and ultimate weight. Unfortunately, their phenomenal growth rate has its disadvantages in that they sometimes suffer from leg and heart problems, which can be fatal. Perhaps the best table bird for the hobbyist is the Sasso, a breed originating in France. It will forage and roost quite happily on free-range and combines hardiness and ease of rearing with a fine flavour. Much useful information can be found on the Sasso company website: www.sasso.fr.

LEFT Some hybrid types may not be exciting to look at but they are excellent layers.

LEFT The Black Rock is a particularly hardy type, a trait that makes it most suitable for keeping free-range.

PURE BREEDS

ONE OF THE GREAT ADVANTAGES THAT PURE-BREED (SOMETIMES KNOWN AS 'PURE-BRED' OR 'STANDARD') CHICKENS HAVE OVER CROSS-BREEDS IS THAT IT IS POSSIBLE TO PREDICT THEIR ADULT SIZE, CONFORMATION AND LIKELY BEHAVIOURAL CHARACTERISTICS, ALL OF WHICH ARE IMPORTANT FACTORS IN DECIDING WHETHER OR NOT THEY WILL BE 'FIT FOR PURPOSE'.

A pure breed is perhaps most easily defined as a true genetic breed – when a male and female of the same breed are mated, they are guaranteed to reproduce a virtually identical type. To refine the definition, a breed is a group of birds that have been produced over a period of many years and nowadays possess inherited characteristics such as shape, colour and comb formation which help to distinguish it from other birds within the same species. Thus, an example of the Ancona breed is very different in size, shape and overall appearance from a Silkie, but both are a species of chicken which is, in turn, a member of the Galliformes order.

RARE BREEDS

If you are intending to exhibit your stock, your chickens will need to be pure breeds, because most show classes cater only for those that conform to a recognized

RIGHT Some pure breeds, such as the Dorking, make delicious table birds.

standard. Even if you have no intention of showing, you might nevertheless prefer to consider pure breeds over hybrids (see previous spread), perhaps because a breed that has taken your fancy is comparatively rare and you have an uncontrollable desire to help improve its status.

However, understanding what is meant by a 'rare breed' is not that straightforward. For instance, in the UK, rather than being rare in overall numbers, a breed might be rare in its geographical concentration, as the Rare Breeds Survival Trust (RBST) points out: 'Some breeds may be numerous but if the majority [of birds] are found in a small geographical area the breed will be highly vulnerable to disease epidemics.' Therefore, their categories of 'critical', 'endangered', 'vulnerable' and 'at risk' are more likely to refer to this factor than total bird numbers.

In the USA, things are very different and the American Livestock Breeds Conservancy (ALBC) Conservation Priority List classifies breeds as rare because of low numbers. 'Critical', for example, means that there are fewer than 500 breeding birds, five or fewer primary breeding flocks, and also that the breed is endangered worldwide. The 'watch' category is used to indicate breeds with fewer than 5,000 breeding examples and also those with genetic concerns or – as with the RBST – limited geographic distribution. Birds classified as 'recovering' include breeds 'which were once listed in another category and have exceeded 'watch' category numbers but are still in need of monitoring'.

CHOOSING YOUR BIRDS

Where unrelated pure breeds are in short supply, it may be that in order to keep certain blood lines and necessary genes going, breeders have had to resort to inbreeding (see Common Breeding Methods, page

141). While this method is often used to fix desirable genes, undesirable genes may also be propagated if it is done without knowledge and understanding. Some strains of Rosecomb bantams, for example, are subject to male infertility as a result of inbreeding. If you are faced with either running this risk or going without, source your stock with extra care and check with the breeder that the birds have not been subject to inbreeding in recent generations.

Good breeder birds are best bought in the autumn when there is likely to be more of a choice. Chickens, pure-breed or not, are generally in shorter supply in the spring; prices will be higher, and the breeds you want may not be available.

Pure breeds are generally more expensive than hybrid birds but conversely, when you have surplus offspring to sell (perhaps as a result of breeding a few additions for your own flock), you may well find that they command a better price. Having said that, no-one should be tempted to breed pure breeds for profit, as the sums often do not add up. It should go without saying that any birds you offer for sale must be sound, free from defects and good examples of their breed for type and colour. Just because you like a particular breed doesn't mean that the rest of the chicken-keeping world shares your enthusiasm and so, rather than having would-be purchasers beating a path to your door, you may find yourself forced to sell at auction. Auction organizers usually charge an entry fee and the auctioneer will take as commission a percentage of the price your birds realize – all of which detracts from your profit margin.

BANTAMS

BANTAMS APPEAL TO YOUNG AND OLD, MALE AND FEMALE, AND CAN BE KEPT VIRTUALLY ANYWHERE. IT IS NO COINCIDENCE THAT EVEN UK CITY DWELLERS WERE ENCOURAGED TO KEEP A FEW DURING WORLD WAR II. HALF A DOZEN CAN BE HOUSED IN THE SPACE REQUIRED FOR TWO OR THREE CHICKENS, THEY CAUSE LITTLE OR NO DAMAGE TO THE GARDEN AND EAT ABOUT HALF AS MUCH FOOD AS LARGE FOWL. THEY MIGHT BE THE SMALLEST MEMBERS OF THE POULTRY WORLD, BUT THEY DEFINITELY POSSESS THE BIGGEST CHARACTERS AND, ONCE THEY HAVE YOU IN THEIR CLUTCHES, YOU ARE A BANTAM FANCIER FOREVER.

Bantams are great 'time-wasters'. Five minutes allocated to feeding and watering can very quickly slip into half an hour as you watch them dust and peck their way across the garden while chirruping in the manner of buxom ladies doing the rounds of market stalls. To watch them busying themselves with their daily activities is wholly therapeutic after a hectic day at work, and just a few minutes in their company helps put life back into perspective. A certain Mrs Ferguson writing in the late 1800s got it exactly right when she said, 'These gems of beauty and most treasured and prettiest of pets are, certainly, the most impudent, as well as diminutive, of our domestic poultry. They are ridiculously consequential, and seem as if they pride themselves on their captivating appearance. There are several varieties, all possessing the same passionate temper and, although such perfect pigmies, are the most pugnacious, which clearly proves their Javanese origin'.

It is often assumed – as Mrs Ferguson did – that bantams are descended from the Bankiva jungle fowl and were given the name by travellers who discovered them in and around the town of Bantam in Java. At first the word was used to describe a particular type, but as the birds began to be exported into Europe, probably some time in the 17th century, 'bantam' became a generic term for any kind of small fowl. As mentioned elsewhere, the modern definition distinguishes between 'true' bantams (for which there is no large breed counterpart) and miniaturized fowl, which are anywhere between one fifth and one quarter of the size of their large breed counterpart. Such clear definitions ought to lessen rather than complicate proceedings – and so they would were it not for the fact that in the US there is a large chicken breed known as the Java and also a true bantam called a Java bantam, and that the two are not related by hereditary bloodlines or genetics.

As children's pets, bantams cannot be bettered. Their character endears them to most people, but they can be particularly rewarding for youngsters, giving them a practical knowledge of biology, livestock management and even discipline, as they begin to realize that bantam keeping is a seven-day-a-week undertaking. Choose a breed very carefully and avoid any that might require specialist knowledge and more complicated housing, because of feathered feet, complicated 'top-knots' or a flighty disposition. It is undoubtedly sensible, especially if you have toddlers, to

RIGHT True bantams, such as the Barbu d'Uccle, right, are naturally small birds and do not have a large fowl counterpart, unlike breeds such as the Welsummer.

opt for one of the heavier, generally friendlier breeds such as the Wyandotte and Silkie. Both are easily tamed, affectionate and rarely show any signs of aggression. They are good layers and often continue producing eggs throughout the winter. The hens make excellent mothers, rarely abandoning the nest before the hatching period is over and being naturally careful when scratching about around their newly hatched offspring. Pekin bantams also are friendly birds that lay reasonably well, come in a variety of colours and, despite their delicate puff-ball look, are quite hardy. In addition, they are just the right size and weight for small hands to cuddle!

A child's natural reaction is to touch and pick things up, so don't expect anything different with livestock. It is important to teach children how to hold bantams correctly. Whereas a large fowl is unlikely to suffer from an unintentionally exuberant squeeze, some of the smaller bantam types such as the Serama might not fare so well.

Although bantams have not been bred for egg-production in the same way as some of the commercial chicken types many breeds can nevertheless match their larger cousins when it comes to the number of eggs laid. Add this to their smaller space requirements and ease of keeping, and they become worthy of serious all-round consideration.

RESCUE & EX-BATTERY BIRDS

FIFTY YEARS AGO, APART FROM SMALL-SCALE PRODUCERS USING A SEMI-FREE-RANGE SYSTEM, ALMOST ALL COMMERCIALLY PRODUCED EGGS CAME FROM HENS KEPT IN THE BATTERY SYSTEM. ACCORDING TO THE THINKING OF THE TIME THEY WERE 'IDEAL WHERE SPACE IS LIMITED...THERE IS NO DENYING THAT PRODUCTION FROM THEM EQUALS, AND VERY OFTEN EXCEEDS, PRODUCTION FROM ANY OTHER TYPES OF ACCOMMODATION'.

Although battery hen types are given names such as Goldlines, ISA Brown's and Warrens, they are all bred specifically to produce the largest number of eggs possible. Interestingly, brown-egg layers are the preferred option for UK egg-producers, whereas in the US it seems that the egg-buying public much prefers white-shelled eggs. There has always been the opportunity to buy ex-battery hens quite cheaply as soon as they come to the end of their first proper laying season. The majority of commercial egg-producers replace them with pullets once the moult has begun. Unfortunately, until relatively recently, few people showed any interest in these birds, even though they had several good egg-laying seasons in front of them, with the result that perfectly good hens were killed off to supply chicken pies and pet foods. Nowadays, thanks mainly to huge efforts made by the Battery Hen Welfare Trust, a charity set up by Jane Howorth in 2004, the ex-battery hens of the UK at least have a chance of an outdoor life once their intensively housed days are over. They are well worth serious consideration by the would-be poultry keeper, but may need a little extra care and management during their re-acclimatization period.

Used only to a small cage, ex-battery birds will probably be poorly feathered, unaccustomed to a nest-box and floor litter in which they can scratch, and total strangers to a perch. Their toenails may also need clipping, although these should quickly shorten once the birds are out and about scratching. In an effort to stop feather-pecking (one or more birds pecking at another), some ex-battery hens will have had their top mandible (beak) trimmed, but there is nothing that can be done about this. However, to make life easier continue feeding dry layers' mash, which almost all commercial setups use. Birds which have been de-beaked will find it far simpler to scoop the mash up with their lower mandible than be forced to peck at pellets. Once the hens have settled it should be possible to gradually introduce other feeds.

The Battery Hen Welfare Trust recommends introducing newly acquired birds to nest-boxes by giving them access to a box on the floor, stating that 'even a cardboard box on its side with shavings and a bit of hay in will suffice until they are fit enough to use the proper facilities'. Because they have no concept of a nest-box, the birds will, in all probability, drop their eggs anywhere, so you should encourage them to lay in the boxes provided by adding artificial or dummy eggs, bought from any agricultural suppliers. Encourage birds to roost by placing perches just a short distance above the shed floor and, if necessary, construct a ramp that will spur them to use the perches at night. Alternatively, lift them all onto the perches once dusk has fallen each evening – they will soon get the idea. When you let them out for the first time, do so a couple of hours before dusk, which will make them more likely to stay near to the house and return as darkness falls.

LEFT With a little care and management ex-battery hens (left) can be given a chance of an outdoor life (right).

BREEDS

THIS SECTION SHOWCASES OVER 70 OF THE MOST INTERESTING AND COLOURFUL POULTRY BREEDS. WHETHER YOU ARE AN EXPERIENCED CHICKEN KEEPER, OR JUST GETTING READY TO CHOOSE YOUR FIRST BIRDS, YOU WILL DISCOVER THAT KNOWING SOMETHING ABOUT EACH OF THE VARIOUS BREEDS IS ALL PART OF THE FUN OF KEEPING CHICKENS.

AMERICAN DOMINIQUE

Sometimes known simply as the Dominique, this was, for a time, the most common backyard fowl in the US. The main reason for the breed's popularity was that it possessed a very hardy constitution and could therefore be kept at little expense by even the poorest of rural families. Heavy plumage protected the birds from the worst of the weather, and their dark and light irregular feather colouring made them practically invisible when scratching in shaded areas helping protect them from predators.

The feathering is barred and very similar to that found in the Plymouth Rock (see page 75). The males are generally lighter in marking than the females, but in both sexes, the light bar should always be wider than the dark. There is no colour variation. The comb when viewed from the top should start rounded at the front before flaring out to the sides and finally rounding back to the spike. The modern bird is dual-purpose, providing meat and laying a medium-sized light brown egg.

ABOVE   Heavy plumage protects the birds from the worst of the weather.

AMROCK

The Amrock originates from the US and is a dual-purpose utility bird that was developed from the Plymouth Rock (see page 75). Nowadays, it is kept mainly by fanciers, who favour it because it can become very tame with careful handling. It is less likely to go broody than some of the other heavy breeds and it lays a good quantity of large brown eggs.

The Amrock is a handsome bird, deep and rounded in shape. Its colouring is similar to that of the Plymouth Rock and Maran (see page 66), but the barred markings are unique in that the cock bird has white and black bars of equal width, while on the hen the black bars are twice as broad as the white ones. The comb is single, and good specimens have yellow legs, although those of the hens will fade to almost white during periods of intense laying.

In the US, prior to the 1920s, any 'barred coloured' birds were known as Plymouth Rocks and were either light or dark in colour. When the US simplified its exhibition standard to accept only the darker type, the genetics found in the lighter barred birds were incorporated into commercial poultry lines. These were given the name 'Amrock' purely as a marketing tool for importing them into Europe. In recent years European breeders have started showing them and breed clubs have been formed in several countries.

ANCONA

Anconas are an ideal beginner's bird but have a tendency towards flightiness. It was once thought they were derived from the Mottled Leghorn and were a member of the Leghorn family (see page 65) rather than a separate breed, but this is now known to be untrue.

Unlike many other modern-day breeds, the Ancona is still much as it was nearly 100 years ago, as evidenced by the following observations in *Fortunes from Eggs*, a book compiled in 1919 by Karswood, producers of a very popular poultry spice of the time:

'It is probably the only breed which has the same type for utility and exhibition, wherefore many of the best layers are bred from exhibition winners. The Crystal Palace Laying Competition was won by Anconas [and] at the Missouri Laying Contest in 1914/15, it was found that, on the average of all pens entered, the Anconas laid 300 eggs for every l00lbs of food [compared to], Leghorns 268, Wyandottes 251, Orpingtons 280, Campines 225, Minorcas 203, Rhode Islands 201, Rocks 188, and Langshans 179. This proves that the Ancona is the most profitable over and above cost of food. Also the Ancona eggs were the heaviest of all breeds. Anconas do well intensively; while on free range they pick up half their own food. They are non-sitters, and grow more quickly than any other breed, while the chicks are easy to rear, and offspring comes true to type and colour. Anconas lay well for three seasons, and can be bred from profitably for four. Use rather light-coloured females [for breeding], not too high in tail with legs well apart, deep "bellies" and a fine, falling comb. Mate to a male with a full flow of feather, a high tail, and a light plumage. Avoid males with thick, fleshy combs and coarse lobes.'

Most commonly black or, more accurately, beetle-green in colour, each feather of an Ancona has a white, v-shaped edge to its tip. Chocolate- and blue-coloured birds also occur and similarly have white tips to every feather. Interestingly, the white tips tend to become bigger after each moult. The comb is usually single, but rose-comb varieties are sometimes seen. Anconas are prolific layers of white eggs.

M

Unlike many other modern-day breeds, the Ancona is still much as it was nearly 100 years ago.

The Ancona originated in Italy and was introduced into Britain in 1851, from where it was exported to other countries, including the US.

APPENZELLER

There are three colours of Appenzeller in Great Britain, but the Spitzhauben – silvery-white, with a black tip at the end of each feather – is by far the most popular. The gold-tipped type should have similar markings to the Spitzhauben (with the obvious difference being in the silver-white and gold 'base' colouring), and there are also blacks, but these are the least common of the three.

The name 'Spitzhauben' means 'pointed hat or bonnet', a possible reference to the traditional Swiss bonnet of its country of origin. Sometimes the breed is also referred to as the 'Pointed Hood'. It has a crest and v-shaped comb. The Appenzeller Barthuhner (known in the US and elsewhere as the 'Bearded Hen') is a more powerfully built bird which lacks the head crest of the Spitzhauben. It also differs in that it has a rose comb with a backward-pointing spike, a full beard and white ear lobes.

In the UK the Appenzeller is a relatively new breed, having been introduced in 1982 by Her Grace the Duchess of Devonshire (now the Dowager Duchess). Despite having its own breed club, it has not really gained many enthusiasts, not least because of its propensity to flightiness, its loathing of confinement and its preference for roosting in trees rather than on the perches of the chicken house. At the time of writing, there are no British standardized Appenzeller bantams, although the bantam is officially recognized in the US.

Each blue feather is surrounded by black lacing, apart from the sickle feathers of the male and the neck hackles of the female, both of which are black.

ANDALUSIAN

This soft-feathered Mediterranean breed originated in Spain – the bird seen in the UK today was developed from black and white stock imported from there in the mid-19th century. By the mid-20th century the Andalusian had become rare, but it seems to have regained its popularity and it is now possible to see specimens at shows in many countries. In the US, it is known as the Andalusian Blue.

Andalusians display colour variations. Each blue feather is surrounded by black lacing, apart from the sickle feathers of the male and the neck hackles of the female, both of which are black. The comb is single and of medium size. The cock bird's comb is well serrated and erect, while that of the female normally flops over to one side. The hen is a good layer of white eggs but, like the Ancona, can be nervy and flighty.

Although a very attractive bird, the Andalusian is definitely not for the beginner, due to its flighty tendencies and the fact that it prefers free-range to confinement. Breeding exhibition birds can also cause problems as there is a good chance that, even if you mate pedigree blue-coloured parents, as many as 50 per cent of the offspring will mature into birds that carry either black or white splashes, which will preclude them from being shown.

ARAUCANA

In most countries, two types of Araucana are seen: tailed and rumpless. The rumpless variety is more akin to the original breed first kept many years ago by the Arauca people of Chile, but today it is the slightly rarer of the two.

A further defining feature of the breed is the tufts of feathers sticking out of the birds' ears. Known as ear rings or ear tufts, these should ideally slant backwards. The tailed variety also has feathered head crests and beards, and both types have a pea comb that is irregular in shape.

The breed contains at least 11 colour varieties. The most popular is lavender, but it also comes in partridge, silver-blue partridge, yellow partridge, fawn, wheaten, white, black and cuckoo. Another unusual characteristic is that the hens lay blue-green eggs (US standards describe the colour as turquoise). Like the Andalusian, the Araucana is a good layer and, being hardy, makes an interesting choice for the novice chicken-keeper.

The Ameraucana is a breed that is sometimes confused with the Araucana. There are several similarities in that both lay a blue-coloured egg, have pea combs, red ear lobes and, when not of the rumpless type of Araucana, ear tufts, head crest and muffs.

The breed's fighting ancestry is still apparent in today's birds and both males and females can be highly aggressive.

ASEEL

Also sometimes spelled Asil, the Aseel was instrumental in the make-up of the Indian Game or Cornish (see page 61). Originating in India, it is reckoned to be one of the purest breeds in the world and is certainly one of the oldest, having been around for at least 2,000 years. It has broad, prominent shoulders, wide hips, a narrow stern and a pea-shaped comb.

The breed should have a moderately low carriage and powerful limbs. The small pea comb, coupled with practically no wattles, is a very desirable point as far as the show bird is concerned and a legacy of the days when they were used for fighting. The minimum amount of head 'furniture' would have given little for an opponent to use as a point of contact. Small combs and wattles also negated the need for the combs to be dubbed.

The breed's fighting ancestry is still apparent in today's birds and both males and females can be highly aggressive: cock birds will frequently fight to the death.

Although known in the US, Aseels are more common in the UK and Europe, where they are bred in various colours of plumage, including black, white, duckwing, dark red, light red, grey and pile. The hens lay tinted eggs and will go broody. Like the Indian Game, they can be difficult to breed successfully, as the cock bird sometimes cannot physically manage the mating process due to his top-heavy build.

AUSTRALORP

The name of this breed is an abbreviation of 'Australian Black Orpington'. It resulted from stock birds imported to the UK from Australia around the 1920s being crossed with the Black Orpington, which was developed by William Cook in the late 1800s (see page 73). Therefore, the Australorp can be said to be both British and Australian in origin.

This soft-feathered bird is heavy and traditionally black, although white- and blue-laced variations do occur. The comb is single, serrated and of medium size, and the eggs are tinted towards brown in colour.

Australorps are easy to keep and become very tame, so they make good pets for children. Unlike some other breeds, they are not aggressive towards one another – even young cockerels being reared as part of a breeding programme can normally be kept together without mishap.

The Australorp is an active breed and, although the birds will happily live in runs, they do enjoy being allowed to free-range. Not being good fliers, they don't need particularly high fencing around a run in order to contain them. For a heavy breed, they are amazingly productive and their

egg-laying performance attracted world attention when in 1922-23 a small flock of six hens set a world record of 1,857 eggs at an average of 309.5 eggs per hen over 365 consecutive days in a trial held in Australia – and that without any of the artificial lighting normally given to commercial breeds during the winter months. A few years later, a new record was famously set when a single hen laid 364 eggs in 365 days.

With such a reputation for domesticity, it is not surprising that Australorps are particularly popular in their 'home' country and that they are well exhibited at shows all around Australia. Australia's standard for Australorps is very similar to the British Poultry Standard, but it is interesting to note that although the white and blue-laced types are known, colours other than black (in reality, a beautiful, iridescent blue-green sheen) have never really taken off in Australia; even though blue-laced bantams are standardized, they are seldom seen on the show circuit.

Australorp fanciers everywhere argue regarding the merits of bantams over large fowl, with many in favour of the bantam type. Reasons include laying a good-sized egg (although they will not lay as many – approximately 170 per season as opposed to up to 260 a year by the standard fowl) and they are cheaper to purchase initially. Some also say that the bantams are hardier than the large fowl, but this is not proven. No matter which size you choose, all types of Australorps make an ideal bird for the beginner.

BARNEVELDER

Imported from the Netherlands to the UK and the US in the early 1920s, the Barnevelder is thought to have Cochin, Langshan and Brahma genes in its make-up (see pages 55, 56 and 54), a mixture that has resulted in a large range of colours. Nowadays, the only colours accepted by most countries' breed standards are black, white, double-laced and blue double-laced, although in Germany blue and brown are also permitted on the show bench.

By far the most popular, and arguably the most attractive, is the double-laced version, which has dark hackle feathers with a bottle-green sheen. However, technically only the females are double-laced and the males in their final plumage are what is more correctly known as melanistic black-breasted reds. The edge markings of the cockerel's feathers appear iridescent green in the light and its tail feathers are black, but on the hen bird the lacing continues right into the tail. All posess a single comb.

The double-laced large fowl Barnevelder originated at the same time as the whole breed was being developed, but the bantam type was not recognized until the 1930s. It was bred in Germany by crossing small large fowl with bantam Rhode Island Reds and Gold-laced Wyandottes (see pages 77 and 88) amongst others. The bantam breed of Barnevelder soon gained worldwide recognition and was exported to many countries because of its ability to lay approximately 180-200 large brown eggs per year. Unfortunately, there has been less emphasis placed on the egg colour in recent years and so, if it is the traditional dark-coloured egg that attracts, it will pay to be very careful when sourcing a particular strain of bantam from a breeder.

As far as the large fowl is concerned, any of the colour varieties would make a good choice for the first-time chicken fancier. They are all easy to keep, either free-range or in a grass run, and, like the Australorp, have an easy-going and placid temperament. In addition, the hens lay a good number of dark brown eggs through the year. Because Barnevelders are not generally well represented on the show bench, this is a good breed for the novice exhibitor, as there is a fair chance of winning some classes during the first season of showing.

The iridescent green sheen on the male's feather makes this a particularly attractive bird.

BARBU D'ANVERS, BARBU D'UCCLE AND BARBU DE WATERMAEL

Collectively known as Belgium bantams, these are all close cousins and are true bantams. Unsurprisingly, given their collective name, they are thought to originate from Belgium, but their country of origin is sometimes given as the Netherlands – a result of their also being found in the Flemish-speaking part of the Netherlands, near the Belgium border.

Belgian bantams were first brought to the UK in 1911 after a display was staged at the Crystal Palace. However, in Australia the Belgium bantam breeds are relative newcomers, having been there only 30 years or so.

Irrespective of type, all have the same colour variations and lay cream-coloured eggs. All are bearded and muffed, but the Barbu d'Anvers has clean legs and a rose comb, whilst the Barbu d'Uccle has a single comb and heavily feathered legs and feet. The Barbu de Watermael has a rose comb and a crest. The feathered legs of the d'Uccle mean these birds generally require a little more attention than some other breeds, especially when considering them for showing and exhibition purposes, but much can be done by keeping them in covered runs with the floors littered with silver sand.

The Barbu d'Anvers is also known as the Belgium d'Anvers or even the Antwerp Belgium. Rumpless versions are given the name Barbu du Grubbe while rumpless d'Uccles are called Barbu d'Everberg. In the US, Belgium bantams are known and prefixed as 'Bearded' rather than 'Barbu'. All are commonly seen as the 'mille fleur' type (meaning 'thousand flowers') which was the original colouring and the first to be imported to the US. Other colours include black, blue, mottled, porcelain, quail, self-blue and white.

As ornamental pets Belgium bantams make attractive additions to the garden, but do not think that because of their diminutive size they will merely potter about looking pretty – they are very active and their light weight and small size make it easy for them to fly into next door's garden. Therefore, it is perhaps advisable to keep them in a closed pen where neighbours may be a problem. The birds will quickly become tame whether confined or given free-range, and their colour variations and ornamental stature make them well worth considering.

> Do not think that because of their diminutive size they will merely potter about looking pretty.

F

BOOTED

Sometimes known as Sabelpoots, Booted bantams are similar in appearance to the Belgium bantams (see previous page) but do not have the beard and muffs; its main characteristic being its quite substantial foot feathering. To maintain these feathers in good condition takes great care; many showing enthusiasts keep the birds either indoors or in enclosed runs with the floors littered with silver sand.

The breed's foot feathering, together with it being more able to flutter than fly, means that any perches incorporated into the housing will need to be wider and lower than for most other true bantam types. Even though they can be difficult to keep, Booteds are nevertheless very popular throughout Europe, possibly because they are less likely to fly than the Belgium breeds. Also, because of their heavy feathering, they are unable to cause damage to the flowerbeds if kept as pets and given free-range of the garden.

Some books give this breed's country of origin as Belgium but, given the fact that they are sometimes known as Dutch Booted bantams, it is more likely that they came initially from the Netherlands. The UK bloodlines were nearly all lost in the early part of the 20th century and it is only through the efforts of dedicated breeders that the Booted exists today – albeit as a rare breed.

BRABANTER

A crested and muffled breed with a v-shaped horn comb, the Brabanter is thought to be Dutch in origin. It is considered more of an ornamental breed than a practical one, and although the hens lay decent-sized white eggs, these tend to be produced in the spring and early summer only.

As an ornamental bird, the Brabanter is appealing for its shape, size and colour rather than its egg-laying or meat-producing characteristics. It is found in black, white, spangled, buff, mahogany, cuckoo and lavender colours, and its crest grows vertically, unlike that of most other crested breeds, where the feathers grow horizontally towards the back of the head. In addition, both male and female have an impressive full throat-beard set between the ear muffs. Of the ornamental breeds, the Brabanter is probably one of the easiest to care for as the birds are good foragers and fairly docile. Since they have small wattles and combs, they are not prone to frostbite, a point worth noting if you are considering keeping them in extreme parts of the US (where they are currently rare). On the downside, they are good at flying and climbing, which may cause problems if you intend to keep them free-range. Fortunately, the breed adapts well to confinement.

The Brabanter is quite closely related to the Owl Beard (see page 74), another ornamental breed, possessing a beautiful upright carriage and sloping, long backline, the only differences being in the crest and the shape of the beard.

BRAHMA

While the most commonly seen Brahma in many countries is the multiple-pencilled partridge or triple-laced variety, other colours include light, dark, Columbian, buff Columbian, cuckoo, birchen, white and gold. Gold cock birds have glossy black breasts, legs and tails, while the hens have delicate black markings on a gold ground colour.

The Brahma originated in India and reached the UK in the mid-1850s, but was recorded in the US roughly a decade earlier. When the birds were first introduced to the UK, they were known as Brahma Poutras and were a great favourite of Prince Albert.

Sometimes known as the 'king of chickens' because of their great size – the cocks can weigh 5kg (11lb) – Brahmas are profusely feathered, staid and matronly in appearance. The feet and legs are heavily feathered and the comb is triple or pea. For a heavy breed, the hens lay a surprising number of light-brown eggs each year but, as is often the case with large varieties, these are relatively small. Unlike many types of poultry, where one cock bird can easily fertilize six or more hens, Brahmas are best kept as trios if you intend breeding them.

CAMPINES

Like the Light Sussex (see page 84), Rhode Island Red (page 77) and a few other well-known breeds, the Gold Campine has been used over the years to produce a reliable sex-linked breed commercially known as the Cambar. As a pure breed, the Campine is seen in two colours: gold and silver. The difference in colouring is most clearly seen in the head and neck feathers, although the gold and silver do permeate through the body feathers, which are predominantly barred with beetle-green.

Although they are good layers, Campines are considered more as ornamental in the US and are few enough in number to be classified as 'rare' in the UK, despite being sufficiently popular to have had their own breed society in the late 19th century.

One of their most notable characteristics is the cock bird's very 'hen-like' shape. It lacks the long, curved tail, saddle and hackle feathers normally seen in males and, at a quick glance, could be mistaken for a hen.

Campines are generally thought to have originated around Antwerp in Belgium. However, Carol Ekarius's recent book *Pocketful of Poultry* suggests that they are descended from the Egyptian Fayoumi, a small breed which has been reared along the banks of the River Nile for many centuries. Mention must also be made here of another Belgian breed, the Braekel, which is similar to the Campine in all respects except size: the Braekel is the larger of the two.

CATALANA

The Buff Catalana is a medium-sized bird and, like the American Dominique (see page 46), is noted for its hardiness. Unfortunately for any would-be fanciers, it is also like the Appenzeller (see page 48) in that it does not take kindly to confinement and should be considered only by those with plenty of land over which it can be given free-range.

The breed is not well known in the UK or US, but it is widely distributed through South America – perhaps unsurprisingly as it originated in Spain (in the district of Prat near Barcelona) and was no doubt taken there by emigrant Spaniards many centuries ago. Unlike many of the Mediterranean breeds, Catalanas are of medium size and therefore come closer to being dual-purpose than most others. Somewhat plain to look at, they are buff with a greenish-black tail and possess a single comb and large, bright red wattles.

As might be expected of a Mediterranean species, these are good layers of cream-tinted eggs and do not often go broody. Despite not being well known in the US, Catalanas are included in the American Standard of Perfection – the official classification of all recognized breeds of poultry. Buff Catalanas are also known as Catalanas del Prat Leonada.

> It should be considered only by those with plenty of land over which it can be given free-range

COCHIN

The earliest examples of this breed began to arrive in the UK from the Far East around 1840. At first the breed assumed different names and it is still sometimes erroneously referred to as the Pekin. In the UK, there is a true bantam known as the Pekin (see page 74) and also a bantam version of the Cochin. However, in other parts of Europe and in the US only the large Cochin and Cochin bantam are known.

These are big birds and could easily compete with the Brahma for the title of 'king of chickens'. The hocks are covered in feathers that unusually curl around the joints, and the legs and feet are also feathered. The comb is small, single and straight, and in particularly good specimens is well serrated. The eggs are light brown and, for a heavy breed, quite numerous. There are several plumage varieties, the most common being black, white and lavender.

As with all feather-legged breeds, Cochins require special attention, and show birds should always be kept on clean, dry flooring. Nevertheless, they do make a good choice for the novice, especially anyone looking for a breed that will reliably go broody.

CRÈVE-COEUR

The Crève-Coeur is one of France's oldest breeds and, like the Bresse, the best known of all French chickens, was originally bred as a table bird. The modern-day type is larger than its ancestors because it was crossed with the Dorking from the UK (see opposite) during the 1800s.

The Crève-Coeur's head is topped by a full crest, in front of which is a horn comb. Like that of the La Flèche (another ancient French breed, see page 64), this forms a pronounced v-shape. Although it is essentially a table bird, the Crève-Coeur lays a substantial number of white eggs. As it can also become hand-tame reasonably quickly, it is a good choice for those who want a breed a little out of the ordinary. However, its crest can attract lice, so it is important to check it regularly and treat any infestation at the first signs.

Another potential problem is that it is not unknown for a Crève-Coeur's crest to freeze in cold weather. Also, the feathering in general may make it unsuitable for locations where inclement weather is the norm rather than the exception.

Spelt as 'Crève-Coeur' in most countries, general references to the breed in the US have the spelling as 'crevecoeur'. In its original French form the name means 'broken heart', presumably an allusion to the shape of the comb.

CROAD LANGSHAN

Originating in China, this breed was imported to the UK by a Major Croad, after whom it is named. It is sometimes called the Black Croad Langshan, as a white strain does exist but this is rarely seen and the black colouring predominates. In Germany, breeders crossed Major Croad's birds with Minorcas and Plymouth Rocks (see pages 67 and 75) in an effort to achieve a more commercial egg-layer, and in doing so developed the separate Langshan breed.

The breed standards classify both Croad Langshans and Langshans as heavy and soft-feathered. Both have a single comb that should have five points, but while the Langshan has clean legs, the Croad Langshan has sparsely feathered ones. The hens lay well and their eggs are light brown to cream in colour.

The original Black Langshans were imported to the US in 1878 and admitted to the standard in 1883. White Langshans were admitted some ten years later. There are now three varieties of Langshans in the US standard: black, white and blue (the last having been accepted as recently as 1987). However, Croad Langhans, as a distinct type are still not recognized in the US, despite being instrumental in the creation of the Jersey Giant (see page 63), along with the Brahma.

Although it can be tricky to source, the Croad Langshan is one of the easier breeds that the newcomer to poultry-keeping might consider. However, some thought must be given to the fact that their feathered legs might be a disadvantage in wet, clay-type soil conditions.

DORKING

Possibly the oldest of English breeds, the Dorking was certainly around at the time of the Roman invasion of Britain in AD 43. Many other breeds – including the Crève-Coeur (see opposite) – were crossed with the Dorking during the 19th century to improve their suitability for the table.

Interestingly, the colour variations of silver-grey, red, white and cuckoo dictate the comb type of the Dorking. Examples of the silver-grey and red should bear a single comb, while the white and cuckoo must possess a rose comb. Other peculiarities of the breed are its five toes and the pronounced boat shape of its body. The hens lay cream-coloured eggs but generally only in season (spring and summer), so cannot be relied upon for a year-round supply. Some strains of the breed are known to suffer from poor feathering and benefit from being rained on, which will keep the feathers in 'tight' condition. However, if you are intending to show your birds this will not be possible, as the sun and rain will cause any light-coloured or white feathers to turn brassy.

As a utility bird, the Dorking can still hold its own amongst the more commercial breeds. Try crossing it with an Indian Game or Cornish (see page 61) to produce an excellent-tasting chicken.

> Some strains of the Dorking breed are known to suffer from poor feathering and benefit from being rained on.

DERBYSHIRE REDCAP

Although it is also known simply as the Redcap, this breed retains a close association with the English county where it originated. It was originally produced as a dual-purpose bird, but enthusiasts today prize it for its rarity and its ability to lay numerous white eggs.

The feathers of the male's neck, hackle and saddle each have a red quill with beetle-green webbing, finely fringed and tipped with black. The back is red, while the breast and underparts are black. The hen is more nut-brown in colour, with a black half-moon spangle at the end of each feather. The tail is black in both sexes. The comb is an accentuated rose in shape and has a long, straight leader.

Redcaps are hardy birds and at their happiest free-ranging. They are terrific foragers and good fliers, as a result of which they require a lot of space. They are long-lived and will continue to lay plenty of eggs for several seasons but, on the downside, they are not very good at sitting their own eggs.

The Derbyshire Redcap Club in the UK has suffered mixed fortunes over the years: it was very active until the late 1930s, then fell into decline. In June 1976 local breeders re-formed the club, and today it is thriving, due no doubt to the resurgence of interest in the breed.

DUTCH

This is another true bantam and is sometimes known as Old Dutch, although the purists say this is an incorrect title. The Dutch Poultry Club of Holland first standardized the Dutch bantam in 1906, and today their standard recognizes over 20 varieties.

Bred from several regional sources, at first the breed was simply known as the Dutch bantam, and very quickly became popular throughout the Netherlands. It is still one of the most common breeds in its native country, is well-liked in the UK and Germany, and is currently gaining status in both South Africa and the US. The first colours were the gold and silver partridge, white and black, with others quickly following, and there are now 13 recognized colours in the UK and at least ten in the US.

Despite only being brought into the UK in the early 1970s, the Dutch bantam has earned a reputation for being an ideal breed for a small area or confined space. Although, like most breeds, the more space you can give it, the better it will fare. They are hardy birds, but they also thrive on loving care and attention and therefore make ideal family pets (seek out strains that are known to be particularly placid). As well as laying a reasonable number of small cream-coloured eggs they are excellent and reliable broodies.

In the US, there is the opportunity for some confusion between the name 'Dutch' and 'Bassett'. In other parts of the world there is a breed known as the Bassette (with 'e'), but in the US Bassett refers to a Bob Bassett of Florida who served as secretary-treasurer of the American Dutch Bantam Society and so 'Bassett bantam' is the nickname of the Dutch bantam.

The proper Bassette (meaning 'small appearance') is quite rare anywhere but Europe and is most frequently seen in the Netherlands and the southern French-speaking part of Belgium. Bassettes are very lively, constantly on the move and quickly become bored, so require large pens and distractions in the form of titbits. Experienced Bassette breeders claim that young cockerels have to be separated from their brood sisters far earlier than other ornamental bantams because they are very quick-maturing and will attempt to mate with the females at an extremely young age.

> They are hardy birds, but they also thrive on loving care and attention and therefore make ideal family pets.

F

FAVEROLLES

The Faverolles originates from northern France and was bred for both eggs and the table. Nowadays, three distinct types are found: the original French, the German and the British. The breed reached the UK in 1886, since when developments in breed standards have meant that the British type carries its tail higher than its German and French cousins.

Despite their heavy appearance Faverolles are good layers, producing creamy-coloured eggs that sometimes tend towards light brown. The breed has five toes like the Dorking (with which it was crossed to produce a larger bird, see page 57), lightly feathered legs, a beard and side muffs. The original colour was known as salmon, and in some countries this is still the only colour permitted on the show bench. However, in Germany and the UK white, black, ermine, blue and Columbian are accepted, while in France the cuckoo colouring is quite common.

Colour variations in the US are restricted to salmon or white. In Australia, the colours permitted are more like those in the UK, including birchen as well as black, buff, cuckoo, ermine, white and blue-laced. The blue-laced is particularly attractive, with each feather edged with a darker shade of the same colour. Attempting to breed some of the more complicated colour variations can be difficult and may require separate pens to produce cocks and hens suitable for the show bench.

FRIESIAN

Classified as a rare breed in the UK by the Rare Poultry Society, the Friesian is a hardy, light, egg-laying bird that originated in the northern parts of the Netherlands. Even the large variety is very small, to the extent that it is not unknown for standard birds to be entered in bantam classes by mistake.

Friesians produce white eggs, are upright and bold looking, and have quite pronounced white lobes. Their legs are slate-blue and the comb is single. Colours can include black, white, red and cuckoo, with the five further varieties – gold, silver, red, lemon and chamois – all being described as pencilled (having numerous small black dots on the feathers of the breast, back and wing shoulders). Although the shape of pencilling varies from breed to breed, in the Friesian the black spots are tiny and teardrop-shaped. As Friesians are light, they are good fliers and therefore it may prove necessary to keep them in an enclosed run to prevent them from roosting in the trees of a neighbour's garden. As with any of the other light, flighty breeds, it is possible to clip the flight feathers of one wing, but the bird could not then be shown until the feathers had regrown during the following autumn's moult.

> As Friesians are light, they are good fliers and therefore it may prove necessary to keep them in an enclosed run.

FRIZZLE

The Frizzle is Asiatic in origin and is generally thought of as an exhibition breed, the bantam variety having always been more popular than its larger relation. It gets its name from its strange feathering, which should curl towards the head and be as tight and even as possible.

In some countries, the word 'frizzle' simply denotes a feather type, but in the UK it is a recognized breed. If you breed from the same strain year after year there is a danger that you will end up with birds that have sparse and weak feathering. To avoid this always use a fresh, well-frizzled male to mate with your pullets.

Frizzles come in a range of colours, the most common being black, white, buff, blue and silver-grey. Other permitted but less common colours include Columbian, duckwing, black-red, brown-red, cuckoo, pile and spangle. Even though it is an exhibition breed, the Frizzle lays reasonably well (the eggs are cream or tinted in colour). However, anyone considering keeping these birds should be aware that their feather formation leaves them unable to cope with particularly wet weather, so they are better housed indoors.

Exhibitors of Frizzles constantly battle the attitude of people who will not treat the breed with the seriousness they feel it deserves, some regarding Frizzles as hybrids rather than a breed in its own right.

It is useful to noth that, as well as being the name of this breed, 'frizzled' can mean a fault in other poultry types and is normally associated with poor plumage.

HAMBURG

Although Hamburgs originated in northern Europe, both spangled and black varieties have been bred for more than 300 years in northern England, where until the mid-19th century they were known as 'Pheasants' and 'Mooneys'. In the UK, Hamburgs are also known as Hollands, whereas in the US Holland is a separate breed. Hamburgs are smart and elegant, with a rose comb and white ear lobes. They lay a reasonable number of white eggs, but are known to be aggressive towards each other if not given sufficient space in the hen-house and run.

Silver Spangled Hamburgs are white with a large black spot at the end of each feather, while Gold Spangled Hamburgs are a rich mahogany colour, with the same black spots.

These attractive birds will enhance any garden, although the spangled varieties may become discoloured or brassy if exposed to too much sun. These feathers will eventually moult out, but any suspicion of brassiness would put paid to a bird's show prospects. Pencilled Gold and Silver Hamburgs are created from separate cock-breeder and pullet-breeder strains through a process known as double mating (see page 31) and is usually carried out to create a particular plumage type or colour. Another example in which double mating is necessary is the Partridge type of the Wyandotte breed (see page 88).

HOUDAN

Introduced to the UK in the mid-1850s from France, where they were first bred as a table bird, Houdans are now used mainly for exhibition purposes, although they are also reasonably good layers of white eggs. Their colours are limited, and although white and lavender are seen in some countries, the black mottled variety is the one most often encountered on the show bench.

The Houdan is heavily crested, probably due to the fact that, generations ago, Crève-Coeur blood (see page 56) was introduced to the breed. The distinctive crest, the complete muffling surrounding the face and the leaf-shaped comb (also called a butterfly) make the Houdan a rarity in the poultry world. To describe the comb further, it is perhaps easier to say that it is in the shape of an oak leaf and looks like two single combs connected only at the point above the beak. As with other crested breeds, the Houdan needs a little extra care in terms of its housing and is best kept indoors or in a covered run. Like the Dorking and Faverolles (see pages 57 and 59), it is unusual in having five toes.

INDIAN GAME

Originally bred in Cornwall for cock-fighting, Indian Game birds were subsequently developed for the table through the introduction of blood from foreign breeds brought into the UK by sailors travelling back from Asia. Because of their origins, they are also known as Cornish Game birds, especially in the US.

Indian Game possess pea combs and red ear lobes. Only three colours are recognized in the UK: dark (chestnut double-laced with greenish black), jubilee (chestnut laced with white) and blue. Elsewhere, a white laced with chestnut is acceptable.

They are chunky in appearance and, with their widespread legs, look a little like a bulldog when viewed from the front. From above and behind, they are wedge-shaped, with the body tapering down to a short but thick-set tail. It is sometimes difficult to breed them, as the cock bird often cannot physically manage the mating process owing to his top-heavy build. The male is unusual in being almost monogamous, and so any breeding pens should contain only pairs or trios.

Despite being bred originally as fighters, nowadays Indian Game birds are good-tempered and can become tame, but they are not good layers, producing only a few cream or light-brown eggs in the late spring/early summer months.

IXWORTH

Compared with many other breeds that have been around for several generations, the Ixworth is a relative newcomer and was first produced between the two world wars.

Taking its name from its Suffolk village of origin, the breed was developed by Reginald Appleyard, who set out to produce a first-class, quick-maturing table breed that could also lay a reasonable number of good-sized eggs. He used several breeds to create the Ixworth, including White Old English Game, Indian Game, White Sussex, White Orpington and White Minorca – all to produce the fine white skin required by the English table market.

Ixworths are quite deceptive to look at, being a great deal heavier and more solid than they appear. They are active, hardy birds which do well on free-range. There is only one variety, white, which has pinkish-white legs and beak, and all-white plumage. The comb is always pea type and the eyes are orange to red in colour.

The breed very nearly died out some 40 years ago, but thankfully a few dedicated and enthusiastic fanciers were on hand to keep things going. The future for this truly English breed is now far more positive than it once was, but it is still very much a rarity. Therefore, should you be tempted to choose Ixworths, make every effort to ensure you are being offered stock with provenance and pedigree rather than birds that are a hybrid approximation of those developed by Appleyard.

Ixworths are quite deceptive to look at, being a great deal heavier and more solid than they appear.

JAPANESE

This is another true bantam breed and perhaps its most noticeable features are the very distinctive, long, upright tail carriage of the cock bird and the short legs of both sexes – so short that they sometimes cannot be seen because they are covered by the thigh feathers.

The breed has been known for many years in Japan, but is a comparative newcomer to Europe where it first appeared in the mid-1800s. Originally, it would have been unusual to see anything but the feather type observed on most 'normal' poultry breeds, but since the 1960s two other feather variations have become increasingly popular amongst Japanese bantam show enthusiasts. The first is the frizzle type, whereby each feather should curl towards the bird's head, and should be as tight and even as possible, whilst the second is similar to the stranded feathering of the Silkie breed (see page 81). The comb of all types should be single and evenly serrated with four or five points. There are at least 17 colours standardized and recognized in the US and almost as many in the UK, although the black-tailed white is the most often seen.

Breeding young birds is not always easy because mating and subsequent fertility can be difficult and the breed also occasionally suffers from a genetic condition that arrests the development of the chick in the egg. Despite these potential difficulties, these pretty and very ornamental bantams are well worth considering by anyone looking for something a little unusual.

JERSEY GIANT

Another bird that was bred exclusively for the table is the Jersey Giant, which has its origins in the US, where the crossing of Brahmas and Croad Langshans (see pages 54 and 56) was instrumental in its development.

When the bird was first exported to Europe in the 1920s, only the black variety was seen, but today it is also possible to find white and blue-laced birds.

Bigger Jersey Giants are usually produced when stock is bred earlier in the season than normal, giving the chicks the best of the summer months in which to grow. In the past it was not unknown for the breed to produce caponized cock birds weighing as much as 9kg (20lb), although modern strains are nowhere near as heavy. At some stage, Australorps (see page 50) were also introduced into the bloodline, the result being that, size apart, the two look remarkably similar. Like most heavy breeds, the Jersey Giant makes a fine pet and is also a good layer of brown eggs. However, it needs a fair amount of space and the hens are not best suited to incubating and brooding their own chicks – although they make reliable mothers, because of their size they are prone to breaking their own eggs and accidentally squashing very young chicks.

In the US, the breed is sometimes known as the Jersey Black Giant – not because of any colouring differences, but simply as an acknowledgement to the Black brothers who developed it in the 1870s.

The Jersey Giant grows into just about the heaviest chicken you can buy, so you will need to consider this when you plan their housing.

LA FLÈCHE

This old breed takes its name from the town of La Flèche in France. The most peculiar aspect of these birds is their strange comb, which is best described as two small, round horns growing upright and parallel to one another, and which has given rise to the breed's alternative names of 'Devil's Head' and 'Satan's Fowl'.

Although the French standard still requires that birds have a crest in contrast to other French poultry breeds such as the Houdan and Crève-Coeur (see pages 61 and 56), the British version of La Flèche has little or nothing in the way of crest, muffs or tassels. These features were bred out by English fanciers, using birds from some Mediterranean breeds to create a smooth head and white ear lobes. Unfortunately, the resultant crosses reduced size and therefore usefulness as a dual-purpose bird. The La Flèche has never acquired a hugely strong following in the US, despite being included in the first Standard in 1874.

The breed's original colour was black, but white, cuckoo and blue-laced varieties are now also accepted by the standards of most countries. As it was first produced as a table bird, La Flèche is a heavy breed. Despite this, the birds fly well and, if given the chance, often prefer to roost in nearby trees rather than a chicken house. Substantial covered runs may also be necessary.

La Flèches are good layers of large white eggs, but in France they are still kept primarily as meat producers.

"The most peculiar aspect of these birds is their strange comb, which is best described as two small, round horns growing upright and parallel to one another."

LAKENVELDER

Originating in the Dutch-German border region (where its name is spelt Lakenfelder), this is a light breed with a single comb. The plumage patterning is a strong combination of black and white; in some books this is described as 'piebald' or 'belted' owing to the bird's white body and, in good specimens, solid black head and tail. The wing feathers should also have some black markings.

The Lakenvelder is not easy to breed to show standards, as there is a tendency for white feathers to appear in the neck – something that would be penalized by a judge.

Lakenvelders prefer to live free-range and can become quite nervy if they are penned in confined spaces. They are good layers of white eggs and only rarely tend towards broodiness. Many keepers claim that their stock prefers to forage for whatever is naturally available rather than accept the hand-outs of regular feeding, but this is not a reason to assume that Lakenvelders can exist on natural food alone.

Due to the breed's scarcity in the US, it is very difficult to find prime examples and consequently virtually impossible to obtain quality breeding stock. The International Poultry Breeders Association (IPBA), based in Patterson, Louisiana, offers an online database for Lakenvelder breeders throughout the world to create and maintain breeding records and ensure the future well-being of many breeds of poultry. Working with established and recognized poultry associations, the IPBA website was launched in August 2008.

LEGHORN

Leghorns have their origins in Italy (Leghorn is the German name for the Italian city of Livorno), but the development of the breed's various forms took place in a number of countries. Large exhibition types are bigger than those bred purely for utilitarian purposes, but even within the exhibition category there are extremes.

All possess a large single comb, but to conform to British standards the comb should flop over, while that of the US and Dutch-German types should be slightly smaller and only partially floppy. Rose-comb types are also allowed, despite being rare. British Leghorns have smaller tails than the Dutch-German large, full-feathered tail and the American spread tail.

The White Leghorn was imported to the UK via the US in the 1870s and the Brown Leghorn followed a few years later. Since then, many other colour variations have been developed. In the UK, black, white, brown, buff, blue and exchequer (a random mix of black and white) are acceptable. In the US, black, white, red, Columbian, partridge, brown, silver partridge and black-tailed red are now regularly seen; while in Europe laced blue, red, yellow, bloodwing silver and cuckoo partridge types are included in the lists of standardized colours. Leghorns lay a plentiful supply of white eggs.

Interestingly, the bantam versions of the Leghorn are considered more docile than the large fowl and may be worthy of consideration where space is limited or if they are being bought as family pets rather than purely egg-producers.

RIGHT In the West the Malay is mainly kept for participation in poultry shows.

MALAY

Another Asian hard-feathered breed, the Malay is also quite rare. Good specimens are tall with a long neck and legs, which are accentuated by the sparse plumage. The comb is compact and walnut-shaped, and the wattles are very small. The breed is described in some standards as having a 'cruel and morose expression', which is an appropriate definition.

This large fowl was first seen in the UK in the early 1800s and became particularly popular among breeders in Devon and Cornwall (possibly because of the need to introduce new blood into the Indian Game, see page 61). The size, shape and character of the Malay are more important than its colour. The shape of the bird is often referred to as 'triple-arched', whereby the first arch is formed by the neck, the second by the high-profile wings and the third by the tail, which is carried folded below the horizontal. The white and spangled colour varieties seem most popular in the UK, but black, pile, red porcelain and duckwing are also found in other countries. The birds produce cream-coloured eggs but are not particularly good layers.

MARAN

Two of the most remarkable qualities of Marans are the dark, chocolate colour of their eggs and that, although a heavy breed, they lay extremely well. However, layers of dark eggs are often bred commercially and so will suffer from the wrong leg colour, badly shaped combs and poor feather markings that exclude them from competitive showing and breeding. The breed originated in France, where nowadays birds often have some feathering on the legs.

Often known as Cuckoo Marans because of their colour, birds of this breed are found in several varieties, including brassy black, black, white, golden, silver cuckoo and Columbian. French fanciers also breed a wheaten-coloured bird. The comb is single and medium-sized. Marans are friendly towards people and members of their own species, so are suitable both as pets and for breeding. The bantam version is ideal for those with only a small area at their disposal, but many bantam fanciers say that some strains fail to meet the depth of egg colour found in the large fowl. According to geneticists, this is because, when compared to the large bird, the bantam's egg has a bigger surface area, but the pigment gland that affects the egg's colouring is smaller. Therefore there is not enough dark pigmentation to entirely coat the shell.

> The knowledge that you are helping to ensure the future well-being of a very rare breed makes the Marsh Daisy worthy of consideration.

MARSH DAISY

The Marsh Daisy was developed around the turn of the 20th century in Lancashire and became a proper, defined breed in 1913. It is believed to have Leghorn, Old English Game, Hamburg, Sicilian Buttercup and Malay in its make-up – resulting in a bird that is somewhat 'gamey' in character and an excellent forager in a free-range situation.

Surprisingly, despite being given sparse attention by its farming owners, the breed was found to lay very well throughout the winter, as well as being capable of producing a bird for the table which possessed an extraordinary amount of good-quality breast meat.

Although the breed is recognized in five distinct colours of black, buff, brown, white and wheaten, the last is the most likely to be seen. In fact, such is the rarity of the breed as a whole, it is unlikely you will come across any colour other than wheaten or brown. Marsh Daisies have distinctive large rose combs with a leader that follows the head, although not as closely as in some other breeds such as the Wyandotte (see page 88). The ear lobes are sometimes a combination of red and white, although the red pigmentation is generally dominant. Slow to mature and with a tendency towards flightiness, the breed's hardiness, reasonable egg-laying ability and the knowledge that you are helping to ensure the future well-being of a very rare breed make it worthy of consideration, provided you can find a reliable breeder from whom to procure stock.

MINORCA

The Minorca is similar in appearance to the Leghorn (see page 65) and, like that breed, is a soft-feathered, light bird from the Mediterranean (although the bantam version is a result of breeding in the UK and Germany). Colours are generally limited to black or white, but blues are standardized in most countries and are sometimes seen on the show bench.

The cock usually carries a large, erect, single comb; that of the hen should fold over. In some countries a rose comb is accepted, although in the UK the rose-comb bantam is not permitted as an exhibition breed. All birds should have large, white, oval-shaped ear lobes, but these sometimes suffer from spots and scabs. As the birds reach old age, the white colouring often spreads to the eyelids and face, making it impossible for them to be used for exhibition. The hens are excellent layers of large white eggs and, as the Minorca is also a pretty breed that needs little special attention, it has much to recommend it.

Although both the large fowl and the bantam are good layers and relatively easy to keep, the bantams in particular can be flighty and may need confining in a covered run. Also, in cold areas, it may be necessary to protect the comb and wattles with petroleum jelly to prevent frost damage.

MODERN GAME

This hard-feathered breed was developed in England as an exhibition bird some time between 1850 and 1900. It was derived from a cross between the Old English Game and the Malay (see pages 71 and 65) by a generation of judges and breeders who were too young to have had contact with gamecocks (cock-fighting was banned in the UK in 1849) and who imagined – erroneously – that a bird with more reach would have had the advantage in a fight.

Therefore, the breeders selected for a longer necked bird, thinking this would look more 'authentic'. With the increasing popularity of showing, judges started looking for taller birds with a shorter hackle and smaller tails. The breed that was developed was originally known as the Exhibition Modern Game, but the word 'exhibition' was later dropped.

The comb of the Modern Game is single and small, and the breed is found in several colour varieties, including birchen, golden birchen, partridge, white bloodwing, silver bloodwing, black, blue and white. Its most notable features are its long legs, elongated neck and overall carriage, which are all important if you intend to try your luck on the show bench. Modern Game birds are not brilliant layers, but because they are small, elegant and soon become very tame, they make excellent pets for the suburban poultry-keeper.

NAINE DU TOURNAISIS

It is said that this Belgian bantam was frequently kept onboard freight ships on the River Scheldt and, as a result, was given the colloquial name of either 'the Captain's Little Chicken' or 'Skipper's Little Chicken'. Originating on the French-Belgian border, at one time it was also known as the Mille Fleurs du Tournaisis and even today it is found only in the mille fleur colouring. However, it is important to note that the colour markings can be irregular and are likely to differ between individuals. In addition, some pencilling may appear on the plumage of the female.

The Naine du Tournaisis ('naine' being the feminine form of the French word for 'dwarf') is clean-feathered and very attractive. The breed is similar in looks to certain types of Old English Game bantams (see page 71), although the wings are slightly more slanted. It has a very alert head carriage, topped off by a moderately-sized single comb.

Although Naines are most commonly seen in Belgium, the Netherlands and northern France, it is possible to source stock in both the UK and the US, and so the breed might be worth considering as an ornamental pet. It will thrive equally well as a free-range bird or confined to a moveable house and run within the garden. In common with other exhibition or ornamental breeds of bantam, it is not a heavy or consistent layer, but what eggs it does lay are a good size in relation to the height and weight of the hen.

> "It will thrive equally well as a free-range bird or confined to a moveable house and run within the garden."

NANKIN

Sometimes called Nankeens, this breed is thought to originate from the Nankin or Nanking region of China. However, counterclaims maintain that it came from the East Indies and was named by traders because of its similarity in colour to a light reddish-yellow cloth that they used to export. Nankins are nowadays considered true bantams, but records show that both large fowl and bantams were originally imported to Britain in the 1700s.

The breed's overall appearance is alert, active and well balanced, the body being medium in length, moderately deep and carried well forward. Either rose or single combs are correct according to the standards of all countries. Sir John Sebright is believed to have used rose-comb Nankins in the development of his namesake breed (see page 79) and the two breeds are indeed comparable in size and stature. The Nankin's leg colouring is described in most standards as 'slate-blue', and the plumage as 'orangey-red to golden buff, with a black tail'. These little birds have a small but enthusiastic following of fanciers, all of whom praise their tameness, incubating ability and mothering skills.

M

RIGHT The appearance of the Nankin is alert, active, and well balanced.

NEW HAMPSHIRE RED

Always a favourite in its native US, but also rapidly gaining popularity in the UK, the New Hampshire Red is thought to have been bred from the Rhode Island Red (see page 77) without the introduction of any other blood. However, they are lighter in colour than Rhode Island Reds, though they too are classed as heavy and soft-feathered. The New Hampshire is probably best accepted as dual-purpose, although it was originally bred as a layer.

Although the large-fowl type originated in the US, the New Hampshire bantam was developed in the Netherlands and should, in theory, be a true miniature of the large version. As well as the most commonly seen colour, chestnut-red, blue-marked and white examples are sometimes seen, although the standards adopted by some countries preclude the showing of any but the original chestnut-red colouring.

New Hampshire Reds are popular competition birds, especially in northern Europe, and are an excellent breed for the novice as a single colour (in this case chestnut-red) is easier to breed to exhibition level. They also have a placid nature and make perfect family pets, and as they are good layers they will keep a family in brown-coloured eggs for a good proportion of the year.

NORFOLK GREY

The Norfolk Grey was created by Fred Myhill of Norwich, and was first shown at the 1920 Dairy Show under its original breed name of 'Black Maria'. Despite an enthusiastic reception, the fact that it was (and still is) a reasonable layer of light-brown eggs and also capable of attaining a good carcass size for meat production, it never really became popular as a dual-purpose bird.

Its fortunes were such that in the 1970s true breeding stock directly descended from those bred by Fred Myhill was reduced to a single quartet. As is often the case, it was only due to the concentrated efforts of a handful of men that the breed is still in existence in the UK today. Still classified as 'rare', the Norfolk Grey nowadays has a dedicated following in both Britain and the US.

Although it is described as a heavy breed, the Norfolk Grey is not that large and weighs quite a bit less than others under the same classification. It has a single comb and is attractively marked. The cock bird has a silver hackle striped with black (as are the neck, back, saddle and wing feathers), while the rest of the plumage is black. The hen's hackle is similar to that of the cockerel, but otherwise she has black feathering all over, apart from the throat, which is silver-laced.

NORTH HOLLAND BLUE

In Great Britain and the US, the legs of the North Holland Blue should be feathered, but in their home country of the Netherlands the breed standards stipulate no feathers on the legs. The bird must always be blue-grey cuckoo in colour, making it very similar in appearance to the most commonly seen Maran colouring (see page 66). This detail can perhaps be best explained because both Marans and North Holland Blues are connected by a common ancestor, Malines or Mechelen Fowl from Belgium, and also share some Plymouth Rock blood (see page 75).

The North Holland Blue cockerel, like the Maran, is generally of a lighter colour than the hen, and it has a medium-sized red comb and red ear lobes. Introduced into the UK in the 1930s by Les Miles, it was considered a dual-purpose bird. Nowadays, although rare, it has a dedicated following of fanciers who appreciate its docile nature and ease of keeping. Excellent as free-range birds, North Holland Blues are known to become overweight if kept confined, a legacy of the days when they were as much prized for their table-bird characteristics as for their egg-laying capabilities. In some countries, the breed is more correctly known as the 'North Holland Fowl'.

LEFT The bird must always be blue-grey cuckoo in colour, making it very similar in appearance to the most commonly seen Maran colouring.

The overall appearance should be of a well-rounded, muscular bird, which is heart-shaped when viewed from above.

OLD ENGLISH GAME

Like the Modern Game (see page 67), the Old English Game has its origins in Britain and was tremendously popular as a fighting bird. Over the years some 30 colours have been known in this large breed, and in several countries any non-defined colour variations can be shown, as the judges are more interested in body shape and muscle tone.

The overall appearance should be of a well-rounded, muscular bird, which is heart-shaped when viewed from above. Rumpless varieties are also seen. Traditionally, the small, single comb of the male birds was dubbed, or trimmed, to prevent injuries when fighting, but this practice is now illegal in many countries.

There is a notable distinction between the feathering of Old English Game being bred in the UK and those seen in the US and Europe. In addition, there are colour complications and differences between the Carlisle and Oxford variations of the breed, perhaps the most obvious being the angle of the back. The Carlisle's is more or less horizontal, whilst the Oxford's is at an angle of about 45 degrees. Until the period between the two world wars, all types were the responsibility of the Old English Game Club, but in the 1930s, the Carlisle and the Oxford Old English Game Clubs were formed as separate entities. The intention was that the Oxford club would look after the type of bird bred initially for fighting and the Carlisle club would concern itself with the exhibition type.

Perhaps the main reasons for the popularity of the Old English Game are the showing and exhibition opportunities the breed offers. Almost every club or poultry organization has separate classes for all of the bird types and many of the colour variations. However, in Australia the organizers of some shows separate birds into weight divisions rather than by colour differences. The Old English Game Club of America looks after the well-being of the breed in general but the preference for showing bantams over large fowl has meant that there is also an Old English Bantam Club of America which exists to promote the bantam breeds and organizes almost 200 shows every year throughout the US.

Although Old English Game birds are aggressive towards each other, their boldness is a positive attribute when they are kept as pets, because they soon become tame. To help them stay fit and active, they are best kept free-range. Old English are probably the best layers of all the hard-feather breeds and produce a cream-coloured egg.

OLD ENGLISH PHEASANT FOWL

Once known as the Yorkshire Pheasant Fowl, the Old English Pheasant Fowl was officially named in 1914, when a specialist breed club was formed. Photographs of birds from the early 1900s show that the modern bird has changed very little.

The breed today retains much of its original utility merits as a good layer of white or slightly tinted eggs (it was used in laying trials during the 1920s and 1930s), whilst producing a plump little table fowl. Like the Marsh Daisy (see page 66), it is a hardy breed, well suited to free-range and, provided it can be protected from predators, needs little in the way of complicated housing. It is probably not a breed for those keeping poultry in a confined space.

The male bird has a moderately-sized red rose comb, almond-shaped white ear lobes, rich bay and mahogany colouring with striped top and laced breast, slate-blue legs and feet, and a graceful carriage.

The hen is the same colouring, with crescent-shaped spangle markings (a quarter-moon-shaped spot of colour) at the end of each feather. The hens lay an average of 160 medium-sized white eggs each season but the pullets are known to be slow in maturing and do not normally come into lay until they are around seven months old. Unusually for a light breed, they will go broody periodically and make good mothers.

ORLOFF

The Orloff is unusual in that it is one of the few poultry breeds to have originated in Russia, although the bantamized form was created in Germany in the early part of the 20th century. In fact, the large breed is now probably less popular amongst fanciers than the dwarf or miniature version.

US show standards recognize three colour varieties, but in the UK four are accepted, the best known of which is the mahogany (classified elsewhere as red porcelain). The Orloff has a walnut comb and a short, thick beak, and it stands tall on yellow legs. Both sexes have a full beard and bushy eyebrows. The birds are soft-feathered and come under the classification of a heavy breed – the large fowl was originally dual-purpose.

Orloff bantams would make interesting childrens' pets as they are very easy to hand-tame and do equally well whether they are allowed free-range or kept confined. When confined, not being fliers, they do not need as high a fence as some other breeds. Although they are friendly towards their owners, the same cannot be said of their attitude towards each other. When you buy birds to make up an unrelated group, there may be a little initial unpleasantness whilst a new 'pecking order' is established.

Perhaps as a result of their Russian origins, Orloffs are known to be hardy and are unfazed by inclement weather conditions. The hens lay a good number of cream-coloured eggs and, when broody, make reliable and caring mothers.

ORPINGTON

The Orpington makes a good beginner's bird and is popular throughout the world. When the black variety was first introduced at the London Dairy Show in October 1886, it produced quite a stir. Within five years it had been joined by white and buff varieties, and such was the demand that other breeders were crossing Croad Langshan and Cochin breeds (see page 56 and 55) and passing them off as Black Orpingtons.

Today, the four self-colours (buff, blue, black and white) are equally attractive and are almost the only variations ever seen in the UK. Elsewhere, birchen, buff-black laced, barred (or cuckoo), Columbian and partridge types are also quite common.

The Orpington lays a good number of light-brown eggs. It has a small single comb, although the black variety sometimes has a rose comb. Its friendly, docile nature makes either the large fowl or the bantam an excellent choice of breed for the first-time chicken-keeper.

Perfect though the large fowl Orpington is as a beginner's bird, the bantam is in a league of its own when it comes to providing the ideal children's pet. Cuddly and attractive, as well as being very easy to tame, there is the added bonus that although the cockerels will crow occasionally, they do not feel compelled to do so all day. Also, their crowing is nowhere near as loud as that of some other breeds, and so it may be possible to keep a breeding pen of birds in built-up areas without incurring the wrath of close neighbours. The birds' conformation means they have little, if any, inclination towards flying or rushing around and so they can be contained in a wire netting surround of about 1m (3ft), provided they are not placed in any danger from predators or wandering dogs. Also of benefit when considering a children's pet is the fact that they will very easily go broody and make excellent and reliable mothers. So it should be possible to collect a clutch of eggs and place the broody hen in a suitable, safe place, giving young children the opportunity to see the whole incubation, hatching and rearing sequence from beginning to end.

However, the fact that the breed is very amenable and docile does have one small disadvantage in that they may be subject to bullying when kept in a flock of mixed breeds.

OWL BEARD

The most obvious difference between the Owl Beard (sometimes written as Owlbeard) and the Brabanter (see page 53), with which it is sometimes confused, is that the Owl Beard has one continuous full-round beard, while the Brabanter has a three-clump beard. Perhaps better known in Europe than it is in the US, the Owl Beard is currently enjoying popularity in the UK as well as something of a revival in the Netherlands and Germany. Originating in the Netherlands, the breed has been known there for several centuries and is believed by some to have been instrumental in the development of the Poland breed (see page 76).

The colour of the beak and legs in a bird depends on its plumage, which is many and varied. There are self-coloureds such as black and white, but many more appear as spangled, pencilled and even laced varieties. In addition, there is what might be best described as a 'sub-type' known as the Moorhead, which has its own colour variations of white, blue, golden and buff – all with a black head. The productivity of an Owl Beard depends very much on its coloration; black or white types are generally considered to lay the largest number of eggs.

Owl Beards do better as free-range birds than in confinement, a factor that might influence their suitability for certain situations. In all other respects this is a hardy breed that is relatively easy to keep.

PEKIN

There is sometimes some confusion between the bantam type of Cochin (see page 55), which has a large poultry counterpart, and the Pekin, which is classified as a true bantam by the Poultry Club of Great Britain. Both were imported from the Far East in the mid-19th century and some Pekin enthusiasts believe that the first birds were stolen from the private collection of the Emperor of China towards the end of the Opium Wars (around 1860). Others believe that today's Pekin is a derivative of the Cochin.

Whatever the truth of its origin, the breed was not accepted in its own right until the Birmingham Poultry Show of 1969 and in the US only the Cochin is known. British standards stipulate that the Pekin should be single-combed, and 'circular in shape when viewed from above, (the) whole outline rounded (with) heavily feathered legs and feet'.

The Pekin is a very attractive breed that looks well in a small garden in which it can be allowed more freedom than some other birds because its heavily-feathered feet prevent it from scratching and doing much damage to the flowerbeds. Conversely, being so heavily feathered makes it less suitable for heavy clay soil and low-lying areas which are prone to wetness.

PLYMOUTH ROCK

This is a US breed, but the modern-day British type is taller and different in shape from birds found in the US. In the UK, the barred (thought to have been a result of breeding with the Scots Grey, see page 79) and buff varieties are perhaps the most commonly seen, but other colours such as partridge, Columbian, buff Columbian and blue-laced are also accepted. In the US the original Plymouth Rock has become known as the Barred Rock, and the breed has subsequently been further divided by colour.

As a point of interest, the barring in the US variety is broader than that seen in the UK. Similarly, in the US the black bars in the hens are broader than the white, while bars of equal width are desired in the cocks. The black should be beetle-green/black in colour and it should not be possible to confuse the birds with the Black Rock, which was developed as a commercial breed and is often found with a tinge of brown in its plumage.

Any of the Plymouth Rock variations would make a good choice for the novice. They are good layers of cream-coloured eggs (described as 'yellowish' in the US) and their placid nature makes them ideal for a relatively confined area, although like most breeds, they appreciate being given free-range. In addition, being too heavy to even attempt flying means that any fencing need not be all that high. That said, should you choose to include a soft netting 'roof' to prevent wild birds getting in, taking food and introducing bird flu (although this is very unlikely) the sides of the pen might need to be higher, purely for your convenience.

Another point in the breed's favour is that the birds are very amenable in each other's company and, provided that they are given sufficient space, it is possible to run more than one cock bird amongst a large group of hens. The hens, in common with most other soft-feathered, heavy breeds, are excellent sitters when broody and even better mothers, making the Plymouth Rock one of the best options for anyone new to chicken keeping and wishing to hatch off chicks without the need for artificial incubators and brooders.

> **Another point in the breed's favour is that the birds are very amenable in each other's company.**

POLAND

The Poland has a huge, ball-like feather crest and a small, sometimes completely absent, horned comb in front. Interestingly, its name is derived from the English word 'polled', rather than from the country of Poland.

As defined in the US, the UK, Germany, the Netherlands and several other countries, the breed is divided into two groups: the White Crested, with normal wattles; and the Paduaner, which is bearded. Nowadays, the colouring of the first group is restricted to black, blue or cuckoo-barred, while the latter group includes gold- and silver-laced, chamois, black, blue and white.

The colour variations and genetics of the breed make the production of future generations very interesting indeed. For example, the White Crested has dominant genes and, if you were to mate it with one of the self-coloured varieties, most of the offspring would follow the White Crested in appearance. Even breeding self-colours is not without complications as, when mating a blue cock to a blue hen, there is the possibility that only half will be the desired colour, the remainder being a mixture of black and/or 'splash'. Mating a 'splash' to a black will, even more confusingly, most likely produce a hatch of black chicks. To achieve a better definition of barring in the cuckoo variety, the colour will be strengthened by the inclusion of a black. Likewise, to deepen the colour of a blue, it might be necessary to use a black from time to time.

The crest is as it is because the feathers emerge from a protuberance at the top of the skull. Its exaggerated nature means that, in many individuals, the bird's vision is restricted and so a little extra care should be taken to ensure that they are not unintentionally frightened by sudden movements. On the plus side, their poor vision keeps them calmer than many other breeds.

To maintain the crests of Polands in good order, the birds should be kept under cover when it is raining and special drinkers should be used. If you have small pens of birds, nipple or aviary-type drinkers are best, but with larger flocks it is not a bad idea to use old-fashioned pottery pigeon drinkers or even dog bowls covered with a screen of wire netting.

While Polands do need special care, novices should not be put off by the potential extra work involved. The hens lay white eggs.

> **A little extra care should be taken to ensure that they are not unintentionally frightened by sudden movements.**

ABOVE The Poland's remarkable crest is a large part of its attraction, although it is also found in a range of appealing colours.

LEFT Rhode Island Reds can lay more eggs than any other pure-breed hen.

RHODE ISLAND RED

Known simply as the Rhode Island in its native USA, this breed is generally given the suffix 'Red' in the UK and in other countries where a white variety is also found. In the red strains the comb can be either single (most commonly seen) or a rose, but in the white a rose comb is considered more correct. Confusingly, in the USA whites and reds are regarded as two different breeds. The red variety is much darker in colour than the New Hampshire Red (see page 69) but is otherwise very similar in stance and formation.

The Rhode Island Red originated as a result of crossing the Red Malay Game and other Asiatic stock, together with the Leghorn. On the other hand, Rhode Island Whites are the result of crosses between Partridge Cochins, White Wyandottes and Rose-Comb White Leghorns. Some of the early breeders believed that the red variety was the better layer of the two colours, but there is no solid foundation to this belief. Somewhat bizarrely, Rhode Island Reds were exported to Europe as dual-purpose birds long before they were given standard recognition in their country of origin and, as a result, their genetic build-up can still be found in many of today's commercial hybrids. The bantam version was developed mainly in Germany and the UK, where they were created by mating large-fowl Rhode Island Reds with other bantam breeds. Like the large fowl, the bantam should conform to the 'brick-shape' appearance – that is, when viewing from the side, it should be possible to place its body into an imaginary square.

Rhode Island Reds are hardy and happy to adapt to almost any healthy surroundings. These highly-recommended birds are also excellent layers of light-brown eggs.

ROSECOMB

The Rosecomb is thought by many to be the oldest bantam breed in the UK and its existence was noted by writers in the latter part of the 1400s. In fact, it is known that King Richard III was a keen fancier of Rosecombs as a result of being given a pen of them by a Lincolnshire innkeeper. Although the breed is believed to have originated from Asia, the original British name of 'Black African' suggests they arrived in the UK via a detour.

Rosecombs are very pretty, proud birds. They are full of character and, being small, make ideal pets in a limited space – they are perfectly happy confined to a moveable house and run. The most common colour is black, but whites can some-times be seen at shows and blues are becoming increasingly popular. In contrast to the limited number of colours seen in the UK, in the US there are at least 20 variations, including barred, brown-red, buff, creole, exchequer, lemon-blue, mottled, silver duckwing and wheaten.

Despite being very easy to keep on a day-to-day basis, Rosecombs can be difficult to breed because some strains are subject to male infertility. Therefore, if you are seriously intent on breeding it will pay to source stock carefully. Try to gain some assurance from the breeder that the birds have not been subject to inbreeding in recent generations – a practice known to encourage hereditary faults.

SCOTS DUMPY

This breed, whose name is sometimes spelt Scots Dumpie, is very old and may even have been brought to Scotland before the Roman invasion of Britain in AD 43. In its cuckoo colouring, the hen could be mistaken for a Maran (see page 66), although the cock bird is much finer and has a different shape altogether.

The Scots Dumpy has noticeably short legs which gives it a distinctive waddling gait quite unlike that of other chickens. The reason for their short legs and heavy body is traditionally attributed to Scottish crofts and smallholdings being surrounded by wild countryside in which predators were rife, so the birds were bred this way to encourage them to stay close to home.

At one time the Dumpy came in a range of colours but owing to a loss of genetic material, only black, white and cuckoo varieties are seen today. Had it not been for a dedicated band of breeders in the 1970s, the breed could have died out altogether. The birds have a single comb and the tail of the cock is full and flowing. The hen is considered to be an excellent broody and mother, and lays cream-coloured eggs.

SCOTS GREY

At one time this breed was known as the Scotch Grey but nowadays it is almost universally referred to as the Scots Grey. It was popular in Scotland until the 1930s when it seems to have lost favour among chicken keepers. The breed standard recognizes only the cuckoo variety and insists it should have a 'single comb' and 'white mottled legs', and be a 'fine, compact, smart bird with well-defined markings'. The Grey resembles the Scots Dumpy (see opposite) and its single comb is similarly accompanied by red ear lobes.

Although it is known as a hardy breed the Scots Grey is not perhaps the best bird for a confined area. Its hardiness means that it requires space to free-range and, according to some breeders, is equally at home roosting in trees as in the chicken-house. These factors are not that surprising when one considers the breed originates from Scottish farmyard birds well used to foraging and fending for themselves.

Sebrights are real characters and quite easy to keep.

SEBRIGHT

Sir John Sebright developed his namesake bantam at the beginning of the 19th century, making it one of the oldest British varieties of 'true' bantam. Somewhat unusually, the cock bird does not have the sickle and hackle feathers that so easily identify the male sex of many other poultry breeds and he is similar to the hen in having either gold or silver lacing throughout the feathering.

Occasionally, because of their genetic make-up, individual males are infertile and the intricate lacing required to comply with show standards may also be difficult to achieve. However, any negative points are far outweighed by Sebrights being real characters and quite easy to keep. Because of their light, flighty nature, they will probably need some form of netted and fenced area, although they will live quite happily in a combined house and run that can be moved around the garden from time to time, making them an ideal choice for the small garden. However, the boiled egg for breakfast might be a rarity, as Sebrights are generally expected to lay only 60–80 creamy-white-coloured eggs each season.

SERAMA

Perhaps more correctly known as the Malaysian Serama, the breed was developed in the Malaysian state of Kelantan and is supposedly the smallest breed of chicken in the world. As an indication of their height and weight, they range from 15–25cm (6–10in) and 225–335g (8–12oz). In the US, where competitions for the breed are becoming quite popular, they are graded into categories according to size rather than type or colour.

The Serama is variably known amongst its enthusiastic and increasing following as either the Arnold Schwarzenegger or the Dolly Parton of the ornamental bantam world due to its proud carriage, pugnacious appearance and heavily protruding breast!

For anyone who has limited space, yet wants to enjoy the experience of chicken keeping, Seramas might be the answer. Their temperament, combined with the fact that a trio can be housed in nothing more than a decent-sized rabbit-hutch, makes them ideal pets. Alternatively, they would be perfect in the bottom of an aviary where they will thrive on the food dropped by other inhabitants. However, they do generally require a warmer environment than the majority of true bantams and may benefit from being kept indoors or with access to a heat source during the winter months (an ordinary light bulb suspended in their house will be sufficient). Seramas have a single comb and come in almost any colour variety. Interestingly, they do not breed true, and parents of one colour may produce chicks of several different variations.

SHAMO

The Shamo has been known for centuries in Japan, but its ancestors came from Siam (now Thailand). Until fairly recently, it was classified as rare by the UK Poultry Club. The fact that it is no longer on that list does not mean that its numbers are growing, merely that, along with several other hard-feathered breeds, it is now protected by the umbrella of a specialist club.

The breed is unusual in the UK in that even exhibition types have no colour standards. However, in Europe a number of recognized colours have been standardized, including black, white, red porcelain, blue-laced, wheaten, and red- and silver-necked blacks. Three different comb types are allowed, all variations on a theme of small, compact, fleshy lumps. Shamo do not have wattles, but instead have dewlapped throats. The skin on the head and throat is smooth on young birds, but becomes thick and wrinkled after about two years of age.

The birds are easy to manage and, because the cocks are not known for crowing (and when they do, they don't make much noise), it is possible to have a breeding pen even where neighbours live in close proximity. Unfortunately, they are not gentle with one another, and if two cockerels were to get together there is a good chance that they would fight to the death. Like Indian Game males (see page 61), Shamo cock birds are virtually monogamous and should be bred in pairs or trios, kept separate in secure, portioned runs.

The Ko-Shamo is a true bantam and not a miniaturized version of the large fowl Shamo, although it is very similar in its overall appearance.

SICILIAN BUTTERCUP

Developed in Sicily, this breed was first seen in the US in 1835, but it was not until about 60 years later that any serious breeding stock was imported. The breed's arrival in the UK is reckoned to have been as a result of importation by a Mrs Colebeck of Yorkshire and the British breed standard was established in 1920.

The Sicilian Buttercup, in common with many other Mediterranean breeds, was used extensively in laying trials during the 1960s, for its ability to produce very high numbers of eggs when fed on the most basic of rations.

The name 'Buttercup' supposedly comes from the bird's very unusually shaped comb, described as 'a cup-shaped crown with a complete circle of medium-sized regular points' and resulting from a genetic variation of two side-by-side single combs that eventually merged at the front and rear. Another notable characteristic of the breed is the vast difference in colouring between the sexes of the gold-spangled variety. Males are a deep orange-red with a black tail, the cape showing some black spangles, whilst hens are golden-buff, with parallel rows of black spangles. There is a silver-spangled variety, but this is seldom seen, and colours such as brown, white and gold-duckwing, recorded when the breed was first imported from Sicily, are nowadays unknown. Ear lobes are red in the UK standard, but white in the US. The leg colouring, like that of the Derbyshire Redcap (see page 57), is willow-green. The Sicilian Buttercup is sometimes erroneously referred to as the Sicilian Flower-Bird.

SILKIE

Silkies are a light breed that arrived in the UK from China sometime in the early 1800s. Since then, they have become extremely popular, particularly as children's pets – there is no doubt that their unusual feathering fascinates youngsters. Colours include white, black, partridge, cuckoo, red-buff and lavender, although not all of these are recognized everywhere.

The comb is known as a mulberry or cushion and is almost circular in shape but preferably broader rather than long. The male has a slightly spiky crest, while that of the female should be short and neat, resembling a powder puff. Some strains also have a beard. Both males and females have an extra toe at the back of each foot. Also typical of the breed is the blue-black skin colour.

Silkies are docile and easy to keep, but because of their feathering they should always have access to a dry run and warm housing. They lay a creamy-brown-coloured egg and can be expected to produce an average of 105 eggs before going broody. The breed is particularly useful for crossing with others to produce the ideal surrogate mother. Some chicken and bantam breeders keep a pen of Silkie crosses to use purely as sitting hens for clutches of eggs from their main-interest breeds which, for a variety of reasons, are not likely to go broody themselves, or make poor mothers when they do. The reason a cross, rather than a pure-bred Silkie, is preferred is that the crosses have all the superb mothering tendencies of the Silkie but lack the fine, cotton-like feathering, so are less likely to strangle the brooding chicks.

SPANISH

Also known as the White-faced Black Spanish, at first glance this Mediterranean breed is similar to the Minorca and Leghorn (see pages 67 and 65), although on closer inspection it's evident it is slightly longer and more sloping along the back. It also has a single comb and large white wattles. Unlike the Minorca and Leghorn, the Spanish must be black if it is to comply with breed standards.

Developed over many years purely as laying birds, the females very rarely, if ever, go broody – so they are ideal for anyone who requires a breed solely to lay eggs. Unfortunately, like many Mediterranean types, the Spanish is quite noisy, flighty and excitable, making it difficult to keep either in a confined space or in areas where neighbours have to be considered. The bantam version is more placid and might be a suitable choice in such situations.

Although the Spanish was developed as a laying bird, some fanciers have, over several generations, selected breeding birds with a view to accentuating the large white ear lobes and facial skin, with the result that this is now considered an ornamental breed in some countries. The white parts are of particular interest in that, although they are present at hatching, it is not until the bird reaches a year old or more that these quite striking features are fully formed.

STYRIAN

The Duchy of Styria lay between southern Austria and northern Slovenia, and it was from here that Styrian poultry originated. Its characteristics suggest a common ancestry with some of the Mediterranean breeds, but although first mentioned as early as the 13th century, it merely existed until the second half of the 19th century, when breeders began to realize its potential as a dual-purpose bird.

However, their enthusiasm was short-lived and, even though the Styrian was given a great deal of emphasis at various important symposiums, by the end of World War I there was very little interest in it. Fortunately, stock was collected so that a preservation programme could begin. This must have been moderately successful as a breed standard was defined in 1930. World War II prevented any further significant progress on improving numbers and standards, and it was not until 1987 that any serious efforts were made.

The hens are quick to mature, lively and quite good layers of medium-sized, ivory-coloured eggs. An adult cock has a small, fine-shaped head, the comb being of medium size, behind which can often be found a tassel of thin tufts of feathers. Four varieties of colour are known: light brown, white, barred and partridge, the last of which is the most common.

ABOVE LEFT The Spanish chicken breed are also known as Clown-faced chickens because they have a white face.

ABOVE RIGHT Close confinement does not suit the Sumatra, so make sure that you confine them in large grassy runs or, better still, give them total free-range.

RIGHT Sultans are wonderful to show because they look so impressive and eccentric. They are also very calm birds.

SULTAN

For anyone looking for something really unusual – and with a royal pedigree to boot – the Sultan is well worth considering. It was, as its name suggests, kept by sultans of the Turkish courts (although none of today's stock is likely to be descended from these birds) and its heavy crest, beard, muffs and heavily feathered legs would certainly make an interesting talking point as it busied itself around the garden.

However, unfortunately the Sultan is ornamental rather than practical. The very features that make it so attractive render it unsuitable for all but the driest ground, though it could be kept in an enclosed house and run – a situation to which it is particularly well suited. A covered run would also prevent individual specimens from becoming 'brassy' as a result of being out in the sunshine or extreme weather. This is an undesirable trait that would prevent a bird from being shown until the next moult (brassiness can be a great problem to the showing enthusiast of any breed that includes white, lavender or other light-coloured birds).

In the US, black, blue or white colouration is recognized, but in the UK only white is acceptable. Another unusual feature of the breed is that, like the utilitarian Dorkings (see page 57), Sultans have five toes. The small comb, which is almost completely covered by the heavy crest in the cock bird, should be of the v-shaped horn variety.

Potential problems involved in the Sultan's keeping ought not deter the would-be fancier, especially as the breed is known for its friendly temperament.

SUMATRA

These are slender, graceful, striking birds with a dark-blue triple or pea comb – this colouring is referred to by some breeders as 'gypsy'. Red combs, faces and wattles are sometimes seen but are not desirable on the show bench. Generally, only two colours are seen – blue and black – but in Germany a white strain was developed by mating a Sumatra with a Yokohama (see page 89). Elsewhere, there is also a black-red variety.

As well as the unusual 'gypsy' colouring of the comb, face and wattles, the Sumatra stands apart from most other breeds by the male bird growing two or more spurs at the back of each leg. The tail feathering is unusual in that, although very luxuriant, the sickle feathers are positioned in such a way that they keep the main body of feathers from trailing on the floor. However this breed would not be suited to wet, muddy conditions and their long tails necessitate large houses with high perches.

If you are tempted to keep Sumatras, bear in mind you should confine them in large grassy runs or, better still, give them total free-range. However, be aware that they may prefer to roost in surrounding trees rather than in their allocated home. Unusually for poultry that originates in Asia, the Sumatra lays white eggs.

SUSSEX

The well-known Sussex, which developed from the Speckled Barn at around the time of the Roman invasion of Britain in AD 43, was originally a table bird but is now dual-purpose. Outbreeding with Cochin blood (see page 55) probably produced the first Light Sussex and as a result the colour of the eggs became tinted and the numbers produced greatly improved.

Some fanciers do not regard the Light Sussex as a true variety, and in a number of countries it is known by its colour, Columbian. The breed also recognizes six other types: speckled, buff, white, silver, brown and red. Cuckoo variety is also allowed in some parts of the world.

The plumage variations of the Sussex can be difficult to achieve to show standard, but that should not put anyone off trying one of the less common colours. Also, note that if the lighter shades are being shown they need to be kept out of sunlight to prevent them turning brassy. The breed has a single comb and a good, solid body, making it full of character and giving it the look of a typical 'farmyard' bird. The hens are excellent layers despite being classified as a heavy breed. Along with the Rhode Island Red (see page 77), the Sussex, and particularly the Light Sussex, has been a favourite amongst both farmers and backyard poultry keepers in the UK for many years. This was particularly so during World War II when it was kept extensively as part of the 'war effort', providing eggs and meat in households that would otherwise have lacked both. However,

originally the Light Sussex was primarily a table bird, so much so that a whole industry grew up around the areas of Tunbridge Wells and Eastbourne to supply the London meat markets on a daily basis. With subsequent breeding, much of the meat-production aspects of the bird's make-up have been lost in favour of a lighter, egg-laying breed and nowadays both hybrid-sized and much larger traditional types are seen. Those kept for exhibition seem to be more frame than bulk and so, should a table bird be required, it is important to acquire the right stock.

Like the Rhode Island Red, the Light Sussex has been widely used in the development of sex-linked hybrids for commercial use. However, do not let this be confused with the oft-quoted belief that it is possible to sex chicks of the Speckled Sussex type by assuming that the cocks will have white wings and the pullets light brown wings, as there is no proven scientific basis for this supposition.

TRANSYLVANIAN NAKED NECK

At first glance, the Transylvanian Naked Neck looks like a bird that has spent its life with its head and neck through the bars of a battery cage, as virtually all the neck and crop area is bare of feathering.

The scientific reason for this is that any reduction in feather mass improves heat dissipation through the naked area, improving tolerance to heat and leading to greater productivity under high ambient temperatures. These were important attributes in the extreme heat of Eastern countries, where the breed was developed to fulfil a need for commercially viable table birds.

When the breed was first introduced into the UK in the 1920s, the media made much of the suggestion that the birds were a result of a mating between chickens and turkeys and for this reason were called Churkeys. Elsewhere, they acquired the name Turken for much the same reason. In the US, the prefix 'Transylvanian' is not used and the birds are known simply as Naked Necks.

Apart from the lack of neck plumage, the breed is similar in shape and conformation to most other utility birds. It has a single, deeply serrated, medium-sized comb and the feathers of the large fowl can be blue, buff, cuckoo, red or white. Leg coloration is dependent on the plumage colouring.

By no stretch of the imagination can this breed be called attractive, but there is no doubt that its extremely unusual appearance has ensured an enthusiastic following in some quarters.

ABOVE The 'naked' neck improved this breed's ability to tolerate the extreme heat of its native lands.

TUZO

Much of what has been written about the Shamo (see page 80) applies to its cousin the Tuzo, sometimes known as the Nankin-Shamo just to confuse matters.

There is some discrepancy as to the breed's true origins as, according to *Cockfighting all over the World*, a weighty tome written in 1928 by Carlos Finsterbusch, the Tuzo was once bred and owned by members of the Japanese royal family and their courtiers. However, the Japan Poultry Society claim that no breed has ever been known as a Tuzo in their country. There is a possibility that small specimens of the Shamo were known colloquially as 'Tuzi', a name that stuck when US soldiers took examples of the breed back to the US after the cessation of their war service in Japan. The Tuzo's appearance in Europe came about as a result of US birds being imported to the Netherlands in the mid-20th century, but it was not until the 1970s that it became known in the UK.

The importation of several very differing strains of the breed into Europe has meant that some examples follow the Shamo in appearance, whilst others are more like the Aseel (see page 49). Typical features include a triple, pea comb, virtually non-existent red wattles, a very upright and aggressive-looking appearance, and very strong thighs and shanks. Plumage coloration is many and varied, including black-red, brown-red, creole, cuckoo, pile, wheaten, and both gold and silver duckwing.

TWENTE

Twente, or Twense, is the Dutch name for a breed that is also known by its German name of Kraienköppe and was developed on the borders of the two countries. Although not well known in the UK, where it is classified as a rare breed, the Twente is popular in its countries of origin.

The birds are good-looking, with a proud, erect stance, not unlike that of the Old English Game (see page 71). The comb is walnut-shaped in the male but hardly exists in the female. Colours are partridge, silver partridge, blue partridge and silver-blue partridge, although in the UK the two varieties known are called gold and silver.

The birds are robust and hardy, and soon become tame towards humans, but unfortunately are sometimes aggressive towards each other. The hens are good layers. Both large-fowl and bantam types lay cream-coloured eggs and the large fowl are known for laying right through winter, unlike some other breeds. The Twente bantam makes a good mother and will successfully brood and hatch a clutch of seven to nine eggs. Twentes are good fliers and so will require high fencing or, better still, a netted-over run. Keeping a breeding pen, which obviously necessitates the inclusion of a cockerel, is not a good idea in a built-up area, as they are noted for their inordinately loud crowing.

> Both large-fowl and bantam types lay cream-coloured eggs and the large fowl are known for laying right through winter, unlike some other breeds.

VORWERK

This breed originated in Hamburg in 1900 and was the result of work by Oskar Vorwerk, who required a bird that would be not only dual-purpose, but also economical to keep and rear. Vorwerks were recognized as a pure breed in Germany in 1919, but much of the original stock was lost during World War II and the breed had to be reconstituted from what birds remained. Vorwerks were not imported into the UK until very late in the 20th century and then only as a result of an interest by a Mrs Wallis of Arundel, West Sussex.

The breed is only found with black-belted markings on a buff-coloured body and, apart from the base colouring, it is often described as being similar in plumage, shape and stance as the Lakenvelder (see page 64). Like the Lakenvelder, the Vorwerk has a single comb and its legs are slate-blue.

In the US, the bantam version of the breed was developed by Wilmar Vorwerk (who, despite having the same surname, is no relation of the original breeder) by crossing and culling a mixture of bantam breeds until a bird that bred true was obtained. The bantam was recognized by the American Bantam Association standards in 1985. Bantams and large fowl alike are quite flighty and require reasonably high fencing to confine them, but in all other respects they are an easy breed to keep.

ABOVE The handsome Vorwerk owes its survival to the commitment of a handful of dedicated fanciers.

WELSUMMER

This Dutch breed takes its name from the village of Welsum and has blood from the Cochin, Wyandotte, Leghorn, Barnevelder and Rhode Island Red breeds in its make-up. In most countries the Welsummer is found in only one colour, which in the male is similar to that of a black-red Old English Game (see page 71) and in the hen to that of a Brown Leghorn (see page 65).

Unlike other black-red birds such as the Brown Leghorn and Partridge Wyandotte (see page 88), which have pure black breasts, the Welsummer has a breast that is a mixture of brown and black, similar to that of a pullet-breeder Partridge Wyandotte. The birds have a single comb.

Compared to other egg-laying breeds, Welsummers are relatively poor layers, but the dark-brown colour of their eggs is comparable to that of the Maran (see page 66), the only difference being that the shell of the Maran is glossy whereas that of a Welsummer is matt. Interestingly, the darkest eggs often come from poor layers, while the better layers produce lighter-coloured eggs. Therefore, if the breeder selects hatching eggs that are deep-chocolate in colour, they may unwittingly also be selecting for low production in the future.

Although no shell colour affects the quality of an egg's contents, the coloration of a Welsummer's egg is such an important feature of the breed it should never be forgotten, and every effort must be made to maintain the dark-chocolate pigmentation. Indeed, the Welsummer Club has a separate egg standard to be used when judging eggs of the breed. When selecting breeding stock purely as a result of the eggs a hen produces, it is as well to bear in mind that, as an individual bird grows older, the eggs she lays may well become paler. Therefore, if as a pullet she laid dark eggs that were typical of the egg standard, she should still be considered. In addition, one should always select breeding males from hens who are known to lay deep-coloured eggs.

The Welsummer is an ideal breed for the newcomer to chicken keeping, as it is very friendly and will thrive equally well whether kept free-range or in a closed run. It is thrifty in its eating habits and very hardy, making it far less likely to succumb to the health problems associated with other poultry breeds.

Bantam versions are more lively and active, so when kept in an enclosed run, they should be kept occupied by means of daily titbits, or by the addition of a turf of grass on which they can peck and scratch.

WYANDOTTE

The soft-feathered, heavy Wyandotte breed originated in the US as the Silver Laced Wyandotte. The feather lacing came about by crossing a Silver Sebright bantam with either a Cochin or a Brahma and then introducing a Silver Spangled Hamburg. The White variety is a sport of the Silver Laced and was developed in the UK before being exported back to the US, where it was further improved to become an excellent layer of white- or cream-coloured eggs.

The breed has at least 14 colour variations in the UK, and over 22 around the world. However, it is not just the variety of colours or the general appearance of the birds that makes them so popular, but rather their ease of keeping. Some of the colour variations do make for breeding problems – for example, the partridge variety is produced through the process of double mating (see page 31), for which separate pens of pullet- and cock-breeding birds are required. Like many other pale-coloured breeds, the White Wyandotte needs protection from the sun if you intend to exhibit it, but even so it is perhaps the best variety for the beginner.

Another of the Wyandotte's great attractions is its rounded, fluffy-looking shape. It is often described as 'a bird of curves' – the curves being very clearly defined, especially in the head, tail and body sections, each of which should fit within an imaginary circle. It has been said that the thickness and density of soft feathers found in the vent area makes mating difficult and that fertility is sometimes affected as a result. However, the majority of experienced Wyandotte fanciers say that this is not the case. However, the dense feathering might cause some concern in respect of keeping the vent area clean of faeces. Although this is rarely a problem in a healthy bird, a close watch should be kept on any build-up in that area as, left unattended, it will quickly become a breeding site for mites and parasites.

During its development and because of the many and varied breeds used before it became standardized, the Wyandotte has had several types of comb. However, the modern bird should only be seen with a medium-sized rose comb, close-fitting to the head and not so wide that it hangs over the eyes. The legs should be clean, thick-fleshed and yellow – although the colouring may be lighter on older laying hens. The standard requirement for yellow legs can cause a problem to the potential breeder of some types – for instance, some black strains can show a preponderance of dark scales rather than yellow. Despite these possible difficulties, the docile, hardy, friendly and attractive Wyandotte cannot be recommended highly enough as the ideal beginner's breed.

'A bird of curves': Wyandottes of both sexes are much softer in outline than most other breeds.

> The docile, hardy, friendly and attractive Wyandotte cannot be recommended highly enough as the ideal beginner's breed.

YOKOHAMA

The long-tailed Yokohama cannot be mistaken for any other breed. In a mature bird the tail can easily reach 60cm (2ft) in length. In Japan (where the breed originated) fanciers keep birds in conditions that prevent them from moulting, to encourage the tail to grow as much as 1m (3ft) each year. Both single and walnut combs are accepted in the breed standards. The former is most often seen in black-red and duckwing colorations, while the latter is more common in red-saddled and white varieties.

To describe how the Yokohama arrived into Europe and beyond is to explain how the breed came to be called Yokohama. In Japan, there never was a breed known as Yokohama; the birds have simply taken their name from the port at which they were first exported by a French missionary called Girad. As well as France, birds found their way into Germany, where they were further developed by the addition of bloodlines from another long-tailed Japanese breed known as the Minohiki, the resultant offspring eventually being taken to the US.

Keeping Yokohama cock birds in prime condition can be difficult. To maintain perfect show specimens you need taller houses and higher perches than normal. The laying capabilities of this light, ornamental breed can be quite varied. Depending on the strain, some hens will lay only 40–50 eggs in a season, while others can average somewhere around 100.

YURLOV CROWER

Like the Orloff (see page 72), the Yurlov Crower originated in Russia. As its name suggests, it was used in crowing competitions which were extremely popular throughout Russia at the beginning of the 20th century and was probably developed from table birds kept by peasants and farmers.

According to records and website information, even as late as 1985 there were only 200 specimens of Yurlov poultry registered throughout Russia. Several of these were kept at government poultry farms, without which the breed would undoubtedly have been in an even more parlous state. Fortunately, the population in Russia and the Ukraine now numbers thousands rather than hundreds, but it is probably due to an interest in the breed by a few German poultry fanciers that the Yurlov Crower has come to Europe – and this only in the last 20 years. Colour variants are black, white, silver, golden, scarlet-red or black with light yellow, and the breed can be seen with either a single or rose comb. Despite being a heavy bird, it is a relatively good layer and must therefore be classed as dual-purpose.

Whether the male's crow differs from that of other breeds because it was bred for use in crowing competitions, or whether its very different crow made it suitable for competitive purposes, is not made clear in contemporary records, but it is a fact that the Yurlov cocks were entered in classes according to their tone.

WHERE & WHEN
to Get Your Chickens

ONCE YOU HAVE DECIDED ON THE BREED AND
TYPE OF CHICKEN YOU PREFER, YOU NEED TO FIND
A SUPPLIER OF GOOD-QUALITY BIRDS. POULTRY
SHOWS AND POULTRY MAGAZINES ARE GOOD
SOURCES OF INFORMATION WHEN IT COMES
TO FINDING CONTACT DETAILS FOR RELIABLE
BREEDERS, BUT WHERE POSSIBLE IT IS BEST TO
OBTAIN YOUR STOCK THROUGH WORD-OF-MOUTH
RECOMMENDATIONS. EVEN SO, BREEDERS ARE
UNLIKELY TO SELL YOU THEIR VERY BEST BIRDS,
AS THEY WILL WANT TO KEEP THESE TO MAINTAIN
THEIR OWN STRAINS.

BUYING AND SOURCING BIRDS

In addition to publications aimed specifically at the poultry keeper, the classified section of your local newspaper may contain advertisements for poultry. Birds are quite often sold as trios (one cock and two hens), but in some countries a numerical code may be used to describe exactly what is on offer. Such an advertisement may include the name of the breed followed by '1–2', denoting one cock and two hens; '1–0', telling you that a single cockerel is on offer; or '0–3', indicating that three hens or pullets are looking for a new home. Do not be confused by numbers like '06' at the end of the advert, as these merely indicate the year in which the chickens were bred rather than the quantity on sale.

If you have set your heart on a particular breed you may have to travel some distance to find it, but with effort and perseverance you will eventually source exactly what you want.

RIGHT Some poultry shows include 'selling' classes, which can be a good way of obtaining new stock.

USEFUL ORGANIZATIONS

From your initial enquiries about local poultry breeders, you should also be able to find out the whereabouts of your local poultry club. In some cases these are general fanciers' societies, combining all manner of fur and feather (rabbits, guinea pigs, pigeons and so on) rather than chickens and bantams specifically, although usually the majority of their members are poultry keepers and breeders.

The addresses of clubs and societies devoted to specific breeds can be found in specialist poultry-keeping magazines or through the national poultry clubs. In the UK, these are the Poultry Club of Great Britain and the Rare Poultry Society (see right), while in the US, the American Bantam Association and American Poultry Association are on hand to answer questions. The breed clubs generally hold shows and exhibitions several times a year, and these are always good venues at which to make specific enquiries about purchasing birds and to gain general advice. Sometimes they also include selling classes, which are perhaps the best way to obtain your desired stock.

Most of the clubs have websites, and in addition to these there are many other poultry-oriented websites that offer chickens for sale, information on breeds and noticeboard links where you can post your requests or simply communicate with other like-minded enthusiasts.

THE AGE TO BUY YOUR BIRDS

If your chosen breed proves difficult to find, or if your finances will not permit the purchase of adult pure-bred birds, you might consider buying a small pen of bantams to use as surrogate mothers to hatch fertile eggs instead. Such birds should be easy to source in your locality, but they must be the right sort: heavy in appearance, placid and reasonable egg-layers. As they go broody (which they certainly will), you can set fertile eggs bought from a known breeder beneath them to hatch. Fertile eggs for hatching in this way can be bought and transported over long distances much more cheaply than mature stock. It might sound highly risky to entrust such a fragile parcel to a postal or private delivery service but, provided the eggs are carefully packed, there are rarely any breakages.

Another option is to buy day-old chicks and rear them using bantams in the same way. Many of the larger poultry breeders hatch their eggs by artificial incubation and are quite happy to sell chicks to chicken fanciers, who can then bring them to maturity either by rearing them under broody hens or with the aid of electric or gas brooders. Where possible, the chicks should be collected in person but, like eggs, day-olds do not seem to suffer unduly by being transported by rail or road. For the hours spent in transit, the chicks can generate enough body heat to keep themselves warm, helped by being carried in specially designed, insulated boxes. The methods of introducing eggs or day-old chicks to a broody hen are described in Breeding Chickens, page 136.

Purchasing adult birds obviously has its advantages. Not least, there is unlikely to be the frustration of rearing young chicks to maturity, only to find they are either nearly all cockerels and therefore unable to supply eggs, or, if showing is your ultimate goal, sub-quality specimens that do not quite conform to the breed standards. Birds bought at point-of-lay or beyond can be placed in the chicken-shed and run immediately, and do not require intermediate housing or extra care. Also, the maintenance costs of adult birds are lower as there is no need to provide heating, chick crumbs or growers' pellets.

A pullet is a female bird that is in its first year and has yet to commence laying, while a hen is generally an older female that has completed her first season of laying. When buying birds, pullets are a better choice than hens as you will obviously gain more eggs from them over the course of their lifetime. Having said that, if eggs are your main objective there is no reason why you should not consider buying hens that have been discarded by commercial set-ups after their first season, as these birds are usually inexpensive and still have plenty of laying potential ahead of them.

When contacting breeders initially, it is important to state whether you want pullets or hens. Likewise, when selecting a male bird, specify either a cockerel (under 18 months) or an older cock bird.

WHEN TO BUY

In addition to selling surplus birds at shows, poultry clubs often organize poultry sales and auctions. When these are held in the early spring, you can be sure that the majority of birds on offer will be around a year old and in peak condition for breeding or just ready for a long life of egg-laying. Sales held in the autumn may contain young birds bred in the same year or older birds that might be past their prime for showing purposes but are still capable of producing eggs or show-quality offspring in the next rearing season.

Generally, at autumn sales it makes sense to buy stock that has been hatched in the spring of the same year. By the time they are sold, such birds can be kept as adults and should lay, if somewhat sporadically (unless they are a commercial breed), throughout the winter months before really getting into their stride the following spring. Show birds of this age will also be at their peak for breeding and exhibition purposes the following year.

If you are not buying at spring and autumn sales, just choose the best birds as and when they are on offer. However, remember that chicks or young birds always do better and grow more vigorously during the summer months, when they have access to fresh grass and more sun on their backs.

LEFT Chicks bought day-old need to be reared under a foster mother or artificial brooder.

ABOVE A clear, bright, well-pigmented eye is a good indicator of healthy stock.

WHAT TO LOOK FOR

By purchasing stock from reputable breeders or at sales organized by a club or society, you can be reasonably confident that only well-kept and healthy birds are on offer. It is, however, very useful to be able to recognize the typical signs of a healthy bird from its overall appearance.

First, the comb and wattles must be bright and waxy. There are a few exceptions to this rule, such as when the breed has an unusually coloured comb (for example, the Silkie) or when a hen has been brooding chicks or is in moult and not laying. Second, the eyes should be clear and bright and the plumage shiny and full. During the summer,

you may notice broken feathers on the back and neck of females in a breeding pen, but this will probably have been caused by the mating cock bird.

Handle any bird you consider buying, as even the breeds classified as light should feel fleshy and well muscled. Check the breastbone – it should be reasonably well covered with flesh and certainly not sharp like a knife blade. Inspect the vent (anal area) for signs of diarrhoea and for lice and mites. The underpart of the wings nearest the body is another area favoured by parasites, so lift up the wings, gently brush back the fine feathers in the opposite direction to their growth and look carefully for movement

– although tiny, fleas, lice and mites are all visible to the naked eye.

Finally, spend some time watching the flock: they should all be busy scratching around, dusting in the earth, feeding or preening. Avoid buying birds from a pen where even just one individual is moping around or is isolated from the rest of the batch, as this could be a sign of disease and at the very least indicates that all is not well. Also look around the pen to ensure the faeces are firm, well formed and, in part, white in colour. Slimy, watery green or yellowish diarrhoea-like droppings are another indicator of ill health.

GETTING YOUR BIRDS HOME

Before setting out to collect your carefully chosen chickens, make sure the hen-house and run are ready for their arrival and that food and water are in place. It pays to ask the breeder in advance what type of feed your new birds are used to and, if possible, to purchase your food from the same manufacturer. Alternatively, ask the breeder if you can take some home so that you can make the changeover to a new supply gradually.

If possible, borrow a proper poultry crate or basket for transporting the birds. Otherwise, arm yourself with a sturdy cardboard box that has plenty of air holes and a decent lid to deter escapees. Removing the flaps at the top of the box and replacing them with a piece of hardboard cut fractionally oversize can make a more secure lid. Construct hinges and fasteners by threading wire or string through small holes drilled in the box and lid. Hay or straw makes a better covering for the base of a box than wood shavings, as it allows the chickens a better grip. If you are intending to buy birds from a public sale, animal welfare officers may be in attendance to check that your box conforms to government regulations regarding livestock transportation.

Once home, keep the chickens confined to their house for a few hours so that they can become accustomed to their new surroundings. When they appear settled (after a few hours in the day or overnight), lift the pop-hole and let them explore the run area. Don't push them out, as it is better to let them make their own way so that they don't become disoriented. If the house has no confining run, leave the birds shut in for 24 hours before giving them free-range.

Make every effort to get acquainted with your chickens straight away. Always move slowly and spend some time crouching at the feeder, so that hunger and interest will ensure they approach you. Before long, they will quite literally be eating out of your hand.

BELOW Transport birds in a strong container that is not too big for its occupants, otherwise they will slide about in transit.

HOUSING

& Cleaning

HOW YOUR CHICKENS ARE HOUSED
DEPENDS ON THE SPACE AVAILABLE, THE
BREED YOU HAVE CHOSEN AND THE NUMBER
OF BIRDS. WHILE THE INITIAL COST OF A
GOOD HOUSING SYSTEM IS HIGH, IT WILL
LAST A LIFETIME OF CHICKEN KEEPING,
PARTICULARLY IF IT IS WELL MAINTAINED.
ONE OPTION IS TO BUILD YOUR OWN UNIT,
WHICH CAN BE TAILOR-MADE TO SUIT YOUR
REQUIREMENTS AND TO FIT A PARTICULAR
PART OF YOUR GARDEN OR SMALLHOLDING.
ALTERNATIVELY, THERE IS A WIDE
RANGE OF COMMERCIALLY PRODUCED
HOUSING UNITS ON THE MARKET, MANY AT
REASONABLE PRICES.

ABOVE The best of all worlds: this unit provides shade in the summer and shelter in wet weather, and allows access to the garden when convenient.

SPACE REQUIREMENTS

The amount of space required for housing your birds depends to some extent on the method of rearing you adopt. If chickens are given access to a large outdoor space, they will spend very little time in the house, using it only on wet days and for roosting or egg-laying. However, if the birds are kept intensively, the combined floor space of their house and run will need to be greater, as they will spend all their time there and will require sufficient room for feeding, roosting, nesting and exercising. Whatever system you choose, the basic premise is that you should give your flock enough room to ensure that the three Hs – health, happiness and hygiene – are all maintained.

If you intend to keep your chickens intensively, a house measuring 1.5 x 2m (5 x 7ft) should be sufficient for six hens. If a run is attached, it will accommodate 12 birds. For these sorts of numbers, a combined house and run area of 5sq m (6sq yd) is adequate, although much will depend on whether you are keeping bantams, light large fowl or any of the extremely heavy breeds such as Croad Langshans or Brahmas.

It is also important to remember that, unless you keep your poultry-keeping operation very small, one house and run is unlikely to be enough. There will be times when extra space will have to be found to accommodate a broody hen and her chicks, a 'resting' cockerel or a sick bird. If you intend exhibiting your stock, you will also need a separate penning shed where birds can be kept clean and 'trained' to show themselves to their best advantage. In addition, a shed or some kind of food store may be required.

TYPES OF HOUSING

If you are keeping just a few birds, a movable combined house and run (sometimes known as an ark or fold unit) is probably your best option. However, you will need sufficient space to keep moving the run on to fresh ground before the area beneath it turns into a mud-bath, something that is particularly important during the wetter months of the year. The alternative to the movable unit is a permanent house and run.

ABOVE AND LEFT With a little ingenuity your chicken house can become an attractive garden feature, while providing shelter and security for your birds.

MOVABLE HOUSING

If you have enough space to move a fold unit on to fresh grass every few days, then this is arguably a better option than a permanent hen-house and probably creates the least amount of work. Some units have exterior-opening nest-boxes and, in a few cases, attached feeders and drinkers that can be accessed from the outside, thereby obviating the need to enter the shelter other than to clean out the floor litter in the roosting compartment periodically. The chickens can be kept confined but still have access to fresh grass, and, if the unit is moved on a regular basis, there is little chance of a build-up of disease. The chickens can still be allowed free-range when you are around, but they are protected from predators and dogs when you are not.

The housing part of a fold unit usually takes up about a third of the overall length, with the remainder forming the run. Traditionally, the units are A-framed, with one door allowing access to the house and another to the run. As an added refinement, the back of the run may be fully or partially covered to provide the birds with shade from the sun and shelter from the rain. Old-fashioned fold units, as once used by the commercial poultry industry, are still sometimes seen and are often equipped with wheels that make moving them easier. However, most modern-day versions merely have a carrying pole (usually running along the apex of the roof if the unit is triangular) or extending handles at each corner. A strong floor is essential. The base of the unit rests on damp grass and so needs to be thick enough to withstand moisture and being dragged on to fresh ground regularly.

If you have only a trio of birds or half a dozen bantams, an oversized rabbit-hutch-type unit is ideal. One-third comprises a wooden-floored roosting and nesting section, while the remaining two-thirds has no floor, giving the occupants access to fresh grass. As only the front of the run section is wired, this type of housing is totally wind- and waterproof and so would be worth considering if you intend to keep light-coloured exhibition birds. Some designs also permit the whole roof to slide back on runners, which makes routine maintenance very easy indeed.

For some of the true bantams or miniaturized breeds, an actual rabbit-hutch might make a good home. However, the birds would need access to a small exercise pen or should periodically be given free-range to roam the garden.

ABOVE AND RIGHT A movable fold unit is the perfect answer in a small garden, as it can be moved every few days, giving the hens regular access to fresh grass while allowing the previous patch to recover. Within larger set-ups the unit can be used as a pen for breeding birds or laying hens, or as a rearing unit for a broody and her chicks.

PERMANENT HOUSING

The Victorians were very keen on poultry keeping and often made a feature of their hen-houses in the same way that dovecotes evolved to become ornamental. Poultry-houses usually took the form of a lean-to along the side of a walled garden or stable yard, and generally consisted of a shed and run, with the roof of the shed continuing over the run section. Separate doors allowed access to either the house or the run, the floor of which was normally constructed of compounded earth. Straw, peat or dried bracken was spread over the floor and grain was sprinkled in this litter to keep the hens busy.

If you have the space and a suitable wall that is not too close to your own house or that of your neighbour, it might be possible to consider a variation on this theme. If you replaced the floor covering with sand, such a set-up would be ideal for some of the heavily leg- and feet-feathered breeds. The covered run also makes it perfect for light- or white-coloured exhibition birds, which become brassy if exposed to the sun or inclement weather.

A large conventional hen-house can often be partitioned to allow several pens of birds to be kept under the same roof. The only disadvantage to this is that a narrow corridor needs to be constructed through the interior of the shed, so that the daily chores can be carried out from the inside, thereby removing the possibility of having two runs – one on either side of the building.

Instead of buying a commercially available hen-house, you could adapt an existing outbuilding. As mentioned previously, a shed measuring 1.5 x 2m (5 x 7ft) is adequate for about six large hens, or up to 12 if it opens on to a decent-sized run. It will need to be light, airy, well ventilated and predator-proof, and the door should open and close easily, as you will need to use it several times each day for access. Brush away any cobwebs on the inside of the shed, wash the windows and, if necessary, paint the walls to make it brighter.

It is important that you take hot summers into account if you are planning to use an existing structure in which to house your birds. Old stone buildings are well insulated from the sun, but more modern sheds may become stifling in the heat of the day.

LEFT AND ABOVE
Pop-holes need to be easily accessible and large enough for your stock to get through, while offering night-time security from predators.

21st-CENTURY HOUSING

Urban chicken keeping is a massive new lifestyle trend, supported by dozens of innovative products including the best-selling 'Eglu' – the coop for truly chic chickens

Long gone are the days when it was only retired men who kept and bred chickens as a hobby. In the UK, chicken keeping traditionally took place at the end of the garden beyond the vegetable patch or, if the individual local council allowed, on a council allotment. Knowledgeable though these individuals were (and without them today's breeding lines and genetics would be much the poorer), they were themselves a 'rare breed', seen only at poultry club and county agricultural shows. Nowadays, chicken keeping is chic and enjoys the patronage of young and old, male and female, country dweller or city slicker, show enthusiast and amateur devotee who wants nothing more than an egg for breakfast. Experts analysing the current trend believe it has much to do with financial hard times and the general move towards self-sufficiency, but it is more simple than that – chicken keeping is easy, pleasurable, relaxing and, once the essentials have been purchased, relatively cheap.

It might be thought that making your own chicken-house and run would prove more cost-effective than buying a unit ready-made, but this is rarely the case. Commercial poultry-housing manufacturers have jigs, professional tools, previous knowledge and access to importers' discounts as a result of buying their timber in bulk, so it will certainly pay to see what they have on offer. Then, of course, there are chicken units not made from wood at all, but instead constructed of heavy-duty plastic, of which the 'Eglu' is perhaps the best known. Although they have their detractors who claim that a plastic house is more easily blown over in the wind and is subject to condensation, they nevertheless, have many advantages over wood constructions, not the least of which must be their hygiene benefits. Their lightness can also be an advantage in that they can be easily moved around the garden or, if necessary, carried round to a neighbour's house to be looked after when you're away from home. At least one maker can supply these combined house and runs as either flat-packs for home assembly, or delivered and installed by a 'chicken chauffeur' who will, if asked, tell the purchaser a little of what they need to know about general chicken keeping. It is also possible for firms to supply birds to go with their units.

Although perhaps slightly more expensive than traditional units, these state-of-the-art versions are definitely easy to clean and can be pressure-washed on a regular basis (just flip up a catch, lift off the top, remove the perches and away you go); whilst the daily chore of clearing the droppings is easily achieved by simply sliding out the floor and emptying its contents onto the compost heap. All housing units are reputedly predator-proof (the runs being classed merely as predator 'resistant') and come in a variety of colours, so there's almost certainly one that will match your house paint or complement your floral blooms! It is obvious from their manufacture that a great deal of thought has been incorporated into their design and such modern housing may prove to be an excellent choice for first-time

RIGHT Unlike wooden houses, the eglu doesn't require any long-term maintenance, just regular cleaning.

chicken keepers with only a small amount of space at their disposal.

In some countries, it is possible to take modern consumer trends a stage further. There are companies who operate a 'try-before-you-buy' policy whereby they supply a house, run, food and birds which are yours for an initial period of six weeks. If, after that time, you decide to continue with the hobby then you pay the full cost, but if you decide that chicken keeping is not for you, the firm will collect the housing unit, birds and any unused food and you lose only an initial deposit.

Make no mistake, chicken keeping has leapt rather than crept into the 21st century and has very definitely joined the computer age. Not only is it possible to buy the latest in housing online, but it's also feasible to visit various poultry-orientated websites and compare prices and products within minutes. Materials used for equipment are forever being improved, so that plastics, for example, are likely to be UV protected and your feeding and drinking utensils far less prone to the effects of summer sunshine or winter frosts. There are even websites that claim to be an 'online chicken community' where you can buy and sell housing and other equipment, seek the names and location of reputable breeders, gain advice from other members, or simply share the pleasures of chicken keeping via their forums.

RIGHT Make sure the pop-hole is large enough for the breed you have chosen to go in and out comfortably.

STRUCTURAL CONSIDERATIONS

Whether you are designing your own hen-house or purchasing a ready-made unit, it is important to bear in mind the requirements of your chosen breed. If you intend to keep a long-tailed breed such as the Yokohama the house needs to be tall enough to contain perches fixed at such a height that the birds' long, flowing tails will not be damaged when they are at roost. Likewise, a house with a pop-hole only 30cm (12in) square might suit bantams but will make it impossible for a large Buff Orpington to exit or enter.

However, nor should houses be bought for the convenience of the chickens alone. For instance, if you have a bad back you will not enjoy having to get down on your hands and knees every time you feed your chickens, collect their eggs or clean out the roosting compartment of a small ark unit. For the same reason, permanent housing must be at least 1.5m (5ft) high to allow easy access for all those important day-to-day chores. A tall building also has the benefit of improved airflow and ventilation, both very important factors in chicken keeping (see right).

The floor of a permanent hen-house must be sufficiently strong and adequately braced to prevent springiness as you walk on it. This is especially important when it is raised off the ground. Raised houses must also be equipped with a gangplank leading up to the pop-hole to prevent possible injuries to the chickens' legs and feet as they come and go.

WINDOWS AND VENTILATION

Good ventilation in the hen-house is vital in preventing respiratory diseases, and can be provided by natural means such as windows and/or a protected open ridge running along the roof apex. It must be possible to open the windows, but if there is any danger that predators or vermin can get through them,

they should be covered with small-mesh wire netting on the inside. Approximately one-fifth of the available wall space should be given to windows.

In the summer, warm, stale air inside the hen-house leads to an increased incidence of disease, so ventilation is crucial. In winter, when temperatures are cooler, ventilation is necessary only to get rid of stale air and prevent the accumulation of ammonia gas (generated from droppings), and usually takes place through the opening of doors and through cracks under the roof eaves. The main problem at this time of year is preventing draughty conditions, something chickens cannot tolerate.

NEST-BOXES

The hen-house should contain some simple nest-boxes, ideally one for every three laying birds. Each box should measure 30cm (12in) square and be around 35cm (14in) high. They may be made in tiers, in which case an alighting perch should be attached to the front of each box.

The boxes should be placed just off the ground and in the darkest part of the house (usually under the windows), as chickens prefer a quiet, dim place in which to lay their eggs. Remove any broken eggs immediately to discourage egg-eating, a habit that, once formed, is very hard to stop (see page 152).

LEFT Straw or hay makes a good liner to nest-boxes, but it will need to be cleaned out regularly and dusted with louse powder.

TYPES OF BEDDING

Whatever materials have been used to build the hen-house floor, a covering of litter is essential. It serves three main purposes: to absorb the birds' droppings; to provide ground-level insulation; and to give the birds plenty of exercise and interest. Types of bedding include:

WOOD SHAVINGS – the most common form of litter, available in compressed bales from suppliers of horse feeds and equipment. If you manage to source your own local supply from a firm making wood-based products, check that the shavings are safe to use (softwoods are best) and that they have not been treated with toxic or irritant chemicals. Because sawdust is fine and made even more so by the birds' constant scratching, it may cause respiratory problems and so should be avoided.

SHREDDED NEWSPAPER – available in bales from most agricultural stockists. This makes perfect bedding material as it is highly absorbent, contains no parasites and can be burnt or added to the compost heap after use. The only disadvantage of newspaper is that it is so lightweight – on a windy day the surrounding area can end up looking as though a ticker-tape parade has just taken place.

STRAW – probably even easier to obtain than shredded newspaper in rural areas, as most farmers will be able to sell you a bale or two after harvest. Wheat straw is best, as it is most durable, but oat straw is a good second choice.

Although chopped straw is more absorbent and is preferred by chickens, like sawdust it can become dusty and affect the birds' respiratory system. You should never use hay, as it mats together into a carpet and has also been known to make birds crop-bound (see page 162). However, it can be used in nest-boxes or when forming a comfortable base for a broody hen to hatch fertile eggs.

LEAVES – collected in the autumn. Birds will happily scratch around in these. At one time, peat was favoured by poultry keepers, but it is very dusty unless kept damp and its collection is not environmentally friendly.

SILVER OR SHARP SAND – may prove to be the best floor covering if you keep a feather-legged breed. Ask your intended stock supplier what they use and go with this recommendation.

Whatever litter you choose, make sure it is not mouldy or damp. Mould can cause serious respiratory diseases, while wet litter is a perfect breeding ground for the *E. coli* bacterium. The litter base should be about 12–15cm (5–6in) deep and needs to be raked over daily to remove any matted portions, such as under the perches and around water drinkers. In addition, a complete renewal of the litter will be necessary every five or six months (see page 112)

Occasionally, new birds try to roost inside nest-boxes, but this must be discouraged or foul nests and dirty eggs will result. When the chickens are first placed in the hen-house, prevent them from roosting in the nest-boxes at night by covering the fronts of the boxes with a board or a curtain of sacking. It is also a good idea to make the tops of the boxes inaccessible at night by covering each with a steeply sloping board.

If you want to get into serious egg production, consider using trap nests, as they are the surest way of telling how well an individual bird is laying. The trap closes when a hen enters the nest to lay, so she cannot get out until she is let out. Combined with the use of leg rings or wing tags to identify the individual birds, trap nests allow a record to be kept of all the eggs each hen lays. The system demands that the house is visited every two to three hours, but for those who have the time the results can be very interesting.

ABOVE RIGHT Perches should be cleaned regularly to prevent infection.
ABOVE Make sure that new birds don't try to roost inside the nest-boxes.

PERCHES

One or two perches running the length of the hen-house will give your chickens or bantams somewhere to roost at night. If you need to install more than one perch, be sure to keep them at the same height or the birds will all try to roost on the higher one.

The thickness and width of a perch depends on whether you intend to keep bantams or large fowl, but should not be so narrow as to cause damage to the birds' breastbones. For bantams, the perch should be about 5cm (2in) wide and 3cm (1in) thick, while large birds require something slightly more substantial at around 8cm (3in) wide and 5cm (2in) thick. In all cases, the top edges must be rounded off. Always try to use planed wood for perches and never use wood that has bark attached. Rough-sawn timber can cause splinters, while a covering of bark provides an ideal home for avian parasites such as red mite (see page 163). Both types of wood are also difficult to scrape clean.

At least 20cm (8in) of perching space should be allocated to each bird, and if a single perch is being used in a large poultry-house it should be supported every 1.8m (6ft) of its length. The height at which the perches are fixed is also very important. Unless you are keeping long-tailed breeds (which require a higher roosting point), perches fixed at around 60cm (2ft) off the floor are about as high as you should consider, although for some of the heavier breeds such as Croad Langshans, 30cm (1ft) is perfect. Perches should not be too high, even for the lighter, flightier breeds, because the pads of the birds' feet can be injured by hard landings each morning. There is also a danger that sudden jarring may upset the hens' reproductive system. Where a droppings board is used, it can be set at around the heights recommended above for perches and the perch then constructed about 15cm (6in) above it.

If you are housing exhibition birds, it is particularly important that their tails are not damaged by constant brushing against the interior wall. The distance between the wall and the intended perch position should therefore take this into account. Perches must also never be attached permanently to the walls of the house. Instead, they should be fitted into slotted grooves, hooked into brackets or built as a free-standing frame. In this way, they can be removed for regular cleaning and disinfecting, and they can also be taken out altogether if the house is required for a broody hen or for young stock.

RUNS AND FENCING

The fencing used to form a run around permanent poultry buildings should be around 2m (6ft 6in) high, not only to contain flighty birds but, more importantly, to keep predators out. The chicken wire most commonly used for this type of fencing has a 19-gauge thickness and 5cm (2in) mesh, and measures 1.8m (6ft) across the roll. The fence posts should be around 4m (13ft) apart. About 30cm (1ft) of the fencing wire must be dug into the ground and, unless the top of the pen is covered with a nylon net (see below), brackets of about 1m (3ft) should be attached to the top of each fence post on the outside. Three strands of barbed wire (or electric wire, for which you will need a fencing unit) can then be affixed quite tightly to each bracket, and this will prevent most predators from climbing into the pen.

A nylon net can be used to form a roof over the run, propped up with stakes to keep it taut, although upturned plastic plant pots will need to be placed on top of each stake or they will eventually rub a hole in the mesh. Such a covering is recommended, as it will not only deter predators but also help prevent the droppings of wild birds falling into the run and so reduce the risk of infection from bird flu (see page 164).

An extremely effective way to reduce disease of a parasitic nature is to have two pens attached to the house. While one pen remains in use, the other can be rested, rotovated and reseeded to break parasite life cycles.

When building a gate to the run, make sure that the method used to fasten it is secure. A central bolt often seems safe, but don't forget that a determined fox or dog pushing and scratching at the base of such a gate could force an entry. Instead, top and bottom latches are the better choice. The gateway must also be wide enough to allow for the passage of a wheelbarrow or any broody coops and runs that may be needed in the future.

Where different breeding pens of birds share the same dividing fence, it is imperative that some sort of screen is in place to prevent the cock birds from seeing each other. Without it, they will spend more time trying to fight one another than mating with their hens, and some serious injuries could result. The easiest and least expensive way of forming this partition is to use one of the many windbreak materials sold by garden centres and horticultural suppliers. Sheets of corrugated metal are another, long-lasting option and, provided they are painted to blend in with the surroundings, will not look obtrusive. If finances allow or if you know of a low-cost source of planking, solid wood partitions can also be used, but they will need to be at least 60cm (2ft) high for bantams and twice that for large fowl.

BELOW Fencing is vital to prevent pests or predators from scratching their way in or the chickens from pushing their way out.

Electric netting can also be used as a means of protection. It is made of woven plastic string that is further interwoven with fine wire. The horizontal wires are electrified and the thick black strand at the base acts as an earth. A fencing unit provides the power and, as long as the vegetation at the base of the fence is kept trimmed or sprayed so that the fence does not short out, electric fencing is quite effective.

If you have a small permanent house with an integral run, a concrete base over which gravel, sand or bark chippings can be laid provides a hygienic floor. Birds kept in such conditions will remain happy and healthy, as long as they are given a regular supply of greenstuffs in addition to their usual diet (see page 130).

POSITIONING THE HOUSE AND RUN

Chickens can tolerate cold weather but dislike draughts; they love warmth but, having no sweat glands, are uncomfortable in extreme heat; and they delight in pecking through friable soil in search of insects but hate getting their wet feet. Therefore, the ideal spot for a run needs to be well drained and surrounded by some form of windbreak, and wherever possible the front of the hen-house should be positioned to face the morning sun but be protected from its hottest rays at lunchtime and in the early afternoon. At least part of the run should provide some shade at all times of the day.

Also bear in mind the run's location in relation to your own house and those of your neighbours, as even the best-maintained sites may encourage flies and a few unpleasant smells during the summer

LEFT Electric netting is perfect for creating temporary runs for your free-range birds.

months. While you will enjoy watching the antics of your birds as they rush importantly about their business, your neighbours may not, so screening the run with trelliswork, hazel hurdles or other structures over which plants can be grown may well be appreciated. However, do be careful not to erect anything too close to the chicken run that could be used by predators as a means of gaining entry.

Do not site a permanent run on heavy ground with a high water-table. If there is no alternative but to place it in a low-lying area, grade the site so that surface water runs away from the building. It is also a good idea to surround the house with a 2m (6ft 6in) apron of gravel or walkway so that it does not become a mud-bath in winter. Fix guttering to large houses so that rain falling on the roof drains away, into either a water butt or a piped trench.

If possible, position the house near existing pathways so the expense of creating new walkways is avoided. Large houses should be raised off the ground so that rats and other vermin cannot take up residence beneath them. The space under a raised house also has the advantage of providing extra room for your chickens and is an ideal place for them to dust.

Where chickens are confined to a wire run, they will soon make their own dust-bath, but a covered structure is very easy to build and will be much appreciated by the birds. All that is required is a small shelter consisting of four posts and a sloping roof about 1m (3ft) high. Turn the ground over inside the structure and leave the chickens to do the rest. You might like to add some fine soil, wood ash or sand to the mixture, but the most important thing is that the area remains dry.

BELOW Chickens like warmth but not too much heat: make sure they always have shade available.

KEEPING THINGS CLEAN

Perhaps one of the saddest sights in a back garden is a neglected and ramshackle poultry house leaning drunkenly in an out-of-the way corner. Attached to the shed is often an inadequate run that, depending on the time of year, is either a sea of mud or a desert of bare soil. Sometimes cluttered with old cabbage stumps, there is almost always an impressive bunch of nettles growing within the plot. None of this is necessary and, provided that you spend a few minutes on the daily routine (see page 114), should never be seen. Whatever housing system you choose (see pages 101–106) remember the old maxim that 'cleanliness is next to godliness' and you will not go far wrong.

Provided that the daily routine is adhered to and the nightly droppings cleared each morning, there should be very little cleaning on a weekly basis. Any greens thrown in from the vegetable garden as a means of preventing boredom and adding variety to the diet should really be removed every day, but human nature being what it is, they are often left to rot and become covered in faeces. Even so, you should make a point of picking them up and taking them to the compost heap at least once a week. At the same time remove and replace the nest-box litter, remembering to add a fresh dusting of flea powder.

If you have a store room near the chicken shed, make sure it contains at least a rudimentary tool box and repair a loose board or tighten a slack screw as soon as it is noticed. It is far easier to bang in an extra roofing-felt nail than wait until a sudden gust of wind necessitates re-covering the whole roof.

It is not just housing and equipment that needs to be kept clean. You don't want any form of bacteria in your eggs, so it is logical to clean them when you collect them. When an egg is laid, it immediately starts to cool and the contents shrink, causing a 'suction' effect on the surface. If the shell is contaminated with bacteria – this is possible even in the cleanest poultry house – some of the bacteria will be brought in through the porous shell and will remain on the outer shell membrane (just inside the shell) for approximately 24 hours. Some may then penetrate both the outer and inner shell membranes and get into the egg itself, causing it to rot or the embryo to die.

Egg white contains a substance known as avidin, which binds iron. Bacteria need iron in order to multiply. Therefore, it is important not to wash eggs in water containing iron (perhaps as a result of rusty pipes) because this will override the avidin defence mechanism. The best method of cleaning eggs is to wash them in a proprietary solution. There are several on the market, all of which should kill bacteria on the shell surface as well as a fair proportion of those that may have lodged on the outer membrane.

RIGHT An access ramp prevents birds having to jump down from a raised house, and possibly injuring their legs and feet.

RIGHT If you can't keep a daily routine, set aside some time each weekend to clean and maintain your chicken house.

CLEANING ROUTINES

Inside the hen-house, it pays to gather droppings regularly – preferably each day but at least once a week. Make it part of your daily routine to remove the debris from the droppings boards below the perches and replace this with a light covering of fresh wood shavings. If you leave the droppings too long, they will become hard and difficult to scrape off. About once a month in the summer and every two months in winter, nest-boxes need to be cleaned out and the bedding replaced.

In addition to the daily removal of droppings, all the floor litter in permanent houses must be removed at least once a year (preferably twice) and the interior thoroughly disinfected. Try to pick a sunny day for this job and lock the birds out of the shed. Remove as many of the fixtures as possible, including perches, droppings boards and nest-boxes, and disinfect these and the interior with a virucide. Your local agricultural supplier will be able to recommend a brand of safe, non-toxic disinfectant that contains an anti-parasitic medication and this should be used both inside the house and on any of the items you've removed.

A knapsack sprayer is more effective than a bucket and brush, as the higher pressure created ensures that the killing agents reach all the nooks and crannies where parasites love to hide. A steam-cleaning pressure washer is perhaps the best means of application; small units can be bought quite inexpensively.

At the same time as you disinfect the house, remove all covering litter from attached runs that have a concrete or compacted earth base. If you do not, there is a danger that parasites and bacteria could immediately be reintroduced to the house via the chickens' feet. If gravel, sand or bark chippings are used as the litter, it may be possible to rake or riddle the droppings and rubbish from the material, then disinfect it and re-use some of it, thereby cutting down on the cost of buying replacement litter. Let the house and fittings dry out completely before re-assembling them and allowing your birds access to their home again.

Being smaller, movable fold units present a less daunting prospect when it comes to cleaning, so it is not too onerous a task to clear them out and disinfect them every six weeks or so. Even then, it is all too easy to let the six weeks become ten, so to avoid slippage keep a note of when the next 'service' is due, either on your kitchen calendar or in your book of poultry notes.

Feeders and drinkers should be kept clean at all times. Attach a small, stiff-bristled brush to your water bucket or watering can so that you can brush and rinse out the drinkers before refilling them with fresh water each morning – this takes no time at all. Feeders will not get too dirty if dry pellets or mash are used, but they will receive their share of droppings. As a result, you will need to clean them out as a regular part of your daily routine. Every few weeks, drop both the drinkers and the feeders into a cleaning solution and give them a thorough scrubbing.

Using an iodine-based disinfectant to control bacteria likely to occur in water drinkers is effective and safe, but be sure to use only approved chemicals and also to check that they will not stain or otherwise damage the drinkers. Very importantly, check that there are no residues left in the drinkers before using any water-given medication, as there is just a chance that the cleaning chemicals will make the medication ineffective or even poisonous.

Do not forget to give incubators, coops, brooders and other equipment not in daily use an annual spring clean. Incubators and brooders should preferably be fumigated both before and after use. You will find fumigation products at an agricultural stockist or advertised in poultry magazines.

MAINTENANCE OF THE HOUSE AND RUN

As with most equipment, the regular maintenance of poultry houses and runs will ensure that they last for years. By replacing a loose board, re-stapling a piece of sagging wire netting, oiling hinges and regularly treating all exposed woodwork with a preservative, you can extend the life of these components almost indefinitely.

Obviously, any wood preservatives must not contain chemical compounds that are likely to be harmful to the birds, so it is as well to read the instructions on the packaging carefully before making a purchase.

With the majority of modern products it is possible to treat the hen-house both inside and out, and then allow the chickens to return home just as soon as the wood is dry.

In addition to checking the hen-house itself, periodically inspect the perimeter of your runs for breaks in the fencing through which a predator could gain access. Although good-quality brands of chicken wire will last for up to 20 years, some inferior ones are made of lower-gauge wire and are not galvanized quite as thoroughly. It is a good idea to keep the fence free of vegetation so that any holes are more easily noticed.

OPPOSITE You should aim to remove droppings and refresh floor litter as often as you can to prevent the spread of disease.

RIGHT Keep your chickens outside while you are cleaning and repairing their house.

SEASONAL CARE

Anyone who keeps chickens needs to develop that indefinable asset, the stock-person's eye, with which some are born and some eventually acquire from experience. Others will, unfortunately, always remain blind. The secret is empathy, observation and the ability to put yourself in the bird's place whilst at the same time not falling into the trap of assuming that its emotions, perceptions and reactions are the same as your own.

Do not worry that just because the day seems cold to you, your birds will be feeling the same – provided that they can find a little shelter from the worst of the winter weather, they will be just fine. Chickens can cope admirably with cold conditions but need to be protected from wind chill. Therefore, it is critical that their house is sheltered from the prevailing wind or, at the very least, that some form of baffle screen is fitted in front of the pop-hole to deflect any direct draughts. During the cold months, your birds may not seem as active as normal and will almost certainly need to store energy in order to stay warm. Winter is also a critical time from a dietary point of view, so make sure that they have sufficient food, while taking care not to overfeed them. A little extra mixed grain in the afternoon's feed should help.

Commercial egg-producers have long realized the need to extend the winter daylight hours make the most of their birds' laying potential. Even if getting the maximum number of eggs is not your top priority, there is no reason why you should not do the same, and not just for egg-production. Artificially lighting the hen-house will encourage your birds to continue laying (a hen in full production is said to need at least 14 hours of light a day) and also give them more time to feed. While it is normal for artificial lighting to be used to extend the time at both ends of the day, it is probably less disorientating for the birds to be given extra light in the morning only (perhaps by means of an automatic timer) than to risk them being caught on the house floor rather than the perch when it's time for 'lights out' in the evening. One light fitting is normally sufficient for the average-sized house, but always take expert safety advice before installing any electrical device in places where wood and other combustible materials are being used.

If you are planning to do any serious breeding in the spring, ensure that the cockerel has been running with his hens for a least a month before fertile eggs are needed for hatching. Sort out any coops and runs that will be required for a broody hen and her chicks and clean them thoroughly. If you are using an incubator and artificial brooder, these will need running up to temperature and checking that all is functioning correctly.

In the summer, make sure that there is

LEFT Put some time aside every day of every season to watch your birds to check that they are happy and healthy.

RIGHT You can use artificial lighting to encourage your hens to continue laying through the winter months.

some form of shade for your birds. If none exists naturally, part of your seasonal care could involve constructing a temporary shelter under which birds can dust-bathe and hide away during the hottest part of the day. If prolonged periods of hot weather are expected, consider setting up a sprinkler system with a water hose and nozzle. Your chickens will appreciate a fine spray and it will certainly help to maintain their correct body temperature (remember that chickens have no sweat glands and find it difficult to keep cool in hot weather). During the late summer and early autumn when the birds are moulting and not laying, it may be worth changing from layers' feed to a 'maintenance' ration (see page 127). This could prove more beneficial in creating the energy and protein levels the birds need to grow a complete set of new feathers.

As well as daily and weekly cleaning routines (see pages 112–115), there are a few jobs that are specific to a particular season. Traditional wooden houses and their attached runs need an annual treatment of preservative, which must of course, be animal-friendly.

RIGHT Be sure to choose a warm spring or autumn day for any cleaning or maintenance that requires the birds being shut out of their house for any length of time.

FEEDING
Chickens

YOU WILL ONLY GET OUT OF YOUR CHICKENS WHAT YOU PUT IN. IF YOU FEED THEM WELL AND REGULARLY, THERE IS NO REASON WHY THEY SHOULD NOT PRODUCE EGGS, MEAT FOR THE TABLE OR A GOOD CROP OF CHICKS. HOWEVER, IF THEY ARE NOT WELL FED, ANY NUTRIENTS THEY RECEIVE WILL BE TAKEN UP KEEPING THEIR BODY FUNCTIONING. FORTUNATELY, FEEDING YOUR BIRDS IS SIMPLE AND RELATIVELY INEXPENSIVE. THEY WILL FIND INSECTS, SEEDS, GRUBS AND FRESH GRASS THEMSELVES IN THEIR RUN, ESPECIALLY IF IT IS A FOLD TYPE AND CAN BE MOVED REGULARLY. TO SUPPLEMENT THIS, THEY NEED TWO MEALS A DAY, THE CONTENTS OF WHICH CAN BE BOUGHT FROM ANY AGRICULTURAL SUPPLIER OR PET SHOP.

WHAT CHICKENS EAT

Chickens are omnivores and, provided they are given the opportunity to do so, will feed on seeds, herbs and leaves, grubs, insects and even small mammals such as mice. All fowl are foragers, with an evolutionary instinct to range and search for food. They have excellent full-colour vision and a highly developed sense of hearing that, thousands of years ago, enabled them to keep track of the rest of the flock while ranging over vast areas of dense foliage in search of food. Even today's highly-domesticated breeds have the same desire to hunt and, in the case of poultry, to scratch in their search for food.

Domestic chickens are typically fed commercially-prepared feeds that include a protein source, vitamins, minerals and fibre. This usually takes the form of crumbs, pellets or mash, although generally all types of poultry (with certain exceptions, such as laying birds) also benefit from the addition of mixed corn to their diet.

Poultry have differing nutritional requirements depending on their age. Young chickens and bantams, for example, need much more in the way of protein when they are growing fast than when they are older and grow more slowly – between, say, 10 and 18 weeks.

THE DIGESTIVE SYSTEM

Chickens have a very simple digestive system. Food is taken via the gullet (oesophagus) directly into the crop, where, by the end of the day, it will often show as a bulge in the chest, just above the breast and below the throat.

The food is softened in the crop – mash and pellets break down almost immediately, whereas it can take several hours for cereals to do the same. The food then passes through a glandular stomach (proventriculus), which predigests it, and

into the gizzard, where it is ground down by muscular action. As birds have no teeth, they would find it nearly impossible to make use of much of their food, especially wholegrains, without the aid of insoluble grit in their gizzard (see page 130).

NUTRITIONAL REQUIREMENTS

Birds need protein for growth, tissue repair and for developing immunity against disease. Fats and carbohydrates are used to provide heat and energy, any surplus being stored as body fat, and fibre is required in small amounts to keep the bowels working. Finally, vitamins and minerals are needed to maintain health, while the latter are also used in bone and eggshell formation.

In the wild, a bird's food contains all the fats, carbohydrates and plant and animal proteins it requires, plus sufficient levels of naturally occurring vitamins and minerals. However, as domesticated birds, chickens no longer have the freedom to make up their own nutritional package, it is important that you offer them a diet that is well balanced, providing all their requirements at each stage of their life.

LEFT Whether you are using a trough or a hopper you will need to allow enough space for all your birds to feed at the same time.

BEST FEEDS

Fresh feed is paramount to keeping healthy chickens. It is very important only to feed enough for the day and not to continually place fresh food on top of stale pellets or mash that may remain in the bottom of the feed hopper. Also, do not let your feed become exposed to the weather; protect it from rain and vermin. If these elementary precautions are carried out, your birds should never suffer from ailments caused by damp and possibly disease/parasite-infected food.

Commercial balanced feeds are available from local feed stores and come in pelleted, crumbed or mash forms (see the following pages for more details). These feeds contain essential vitamins and minerals and are available for all types and ages of poultry. The fact that they are 'balanced' is crucial; a great deal of time, effort and money has been spent on developing these products so that they provide all the nutrients a bird needs. So unless you are a nutritional chemist capable of producing your own formulation, it makes sense to use only the best feed in order to get the best results in terms of overall health, egg-laying, fertility and show condition.

Feed carefully. Whilst it is important that your birds are given a top-quality feed that is high in proteins and fats, unlimited access could well cause them to become obese. As well as affecting a bird's general health, obesity can lead to problems with egg-laying and fertility. Breeds which prefer a sedentary lifestyle, usually the heavier, placid types rather than the more active Mediterranean varieties, are perhaps more prone to such problems. Try feeding mash rather than pellets, which are easier to peck at and therefore fill the crop more quickly. A plentiful and regular supply of greenstuffs hung from a frame or hook set just above the birds' head height will give them a little extra exercise as they will have to stand on tip-toe or jump up to reach them, however, the most obvious solution is to feed them less!

A chicken's regular, commercially-available diet will always benefit from the addition of greenstuffs (see page 130) and probably the best regime combines feeding them regularly and also allowing them to free-range and pick up odd titbits for themselves. Although chickens always seem to be busy pecking at bits of grass, seeds and insects, these additions to their diet form only a relatively small (but vital) part of their daily intake. A chicken's gizzard is just the right size to cope with this sort of variety. However, if you give too much grass or other greenstuffs, without the birds having to work for it, the gizzard may be unable to cope and a blockage might occur. This problem tends to be more prevalent with rescued commercial intensive or battery birds that have never before seen the light of day and have been fed entirely on layers' mash. Their gizzard is smaller because it has not had to work as hard as one developed on free-range. It will therefore find it difficult to break down too much roughage and when 'challenged' could well become blocked – with quite serious effects.

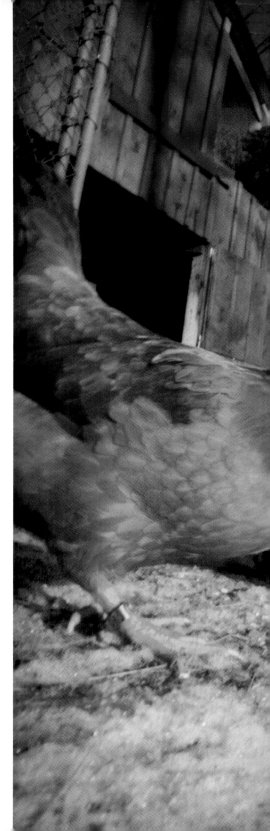

RIGHT Scattering grain for the afternoon feed will keep your birds satisfied throughout the night.

Cereal grains can also be fed along with a commercial balanced ration and most chicken keepers like to feed it in the late afternoon or early evening as it remains in the crop for longer and will keep birds feeling full and warm overnight.

Be judicious when feeding titbits to your birds and remember that they are a treat rather than a part of a balanced diet. With the almost daily creation of new poultry forum websites, it is unbelievable what treats are currently being suggested to first-time chicken fanciers by other (often equally inexperienced) forum members. Many of these are wholly inappropriate. A recent one recommended pineapple chunks; feed those too often and your birds will soon be looking under the weather and excreting a sticky yellow discharge. Beware of feeding any titbits that might ferment in the bird's crop and/or gizzard. These include raisins or stale fruit cake that contains dried fruits – personal experience tells us that the latter can be fatal to bantams. While it is true that chickens and bantams are omnivores and thrive on a varied diet, do not take things to extremes – your birds will always do best on a balanced daily ration of pellets or mash, augmented by tasty little morsels of greenstuffs, cereals and *sensible* titbits.

ABOVE Chick crumbs can usually be phased out at around three or four weeks, but some inbred strains or delicate true bantams may require longer.

BROODY HENS

A broody hen eats very little when she is sitting, and requires not much more than a simple diet of mixed corn, with plenty of maize to maintain body mass and warmth, along with fresh drinking water. No other specific nutrients are required. If fed grain she will also produce firmer droppings, making her less likely to foul her nest.

CHICKS

Even if you have a home-produced or organic feeding programme in mind, do not be tempted to rear young chicks on household scraps. Without the correct protein levels at this stage, they are likely to die or become undernourished. Proprietary chick crumbs are essential, because they contain not only all the nutritional requirements, but also some necessary medicinal additives (for example to combat coccidiosis, – see page 165).

GROWERS

Proprietary growers' rations are the most sensible option for growing birds. They have a lower protein and vitamin content than chick foods. Pullets must not be overfed or given too much protein, or they will grow too quickly while still being immature internally.

This will cause various egg-laying problems and possibly a partial moult at point-of-lay (see page 160).

LAYERS

Buy a good balanced layers' ration that provides around 17 per cent protein, and feed it either as a coarse-ground mash or pellets. Grain and other low-protein foods should not be given with layers' feeds, as their addition will unbalance the manufacturer's carefully evolved formula, although in practice this appears to do little or no harm to the small, non-commercial flock.

TABLE BIRDS

If you are raising surplus cockerels for the table, household scraps can be an invaluable addition to their diet when combined with a high-protein proprietary fattening meal, and will certainly help in bulking them up. A bird fed in this way will have a great deal more taste than the majority of commercially-produced chickens, and its varied diet (especially if maize is given) will make the carcass virtually self-basting. With this method of feeding, a bird is ready not after a certain number of weeks, but when it has put on enough bulk – pick it up, assess its weight and feel its breast. Cockerels as young as ten weeks can make magnificent eating; in the catering trade these are known as poussins and command a high price.

BREEDING BIRDS

Birds that are used for breeding need a plentiful supply of protein and trace elements in their diet or their offspring may suffer from deficiency-related diseases. Breeders' rations are suitable for inclusion in the diet of birds you are intending to breed next year's stock from and should be fed from mid-winter onwards. Maintenance rations can be fed to breeding birds, and can also be used during any period of a bird's life when its body is just ticking over, such as when it is in moult. Consider buying breeders' and maintenance rations produced for game birds, as these are some of the best feeds available.

COMMERCIAL FEEDS

Assessing and combining all the required ingredients of a chicken's diet is a fine art, but fortunately there is no need to produce a perfect home-made mix now that balanced commercial foods are available. Anyone worried about unknown additives in commercial poultry foods can choose from free-range and organic rations that are free from such chemicals. Whatever your choice, buy high-quality foodstuffs from reputable firms.

Ornamental-fowl pellets are available from some of the more specialist food suppliers. They are equally useful for bantams or large fowl, and are especially valuable if you want your birds in prime condition immediately prior to the show season. The pellets are also smaller than regular pellets, making them ideal for bantams.

PELLETS OR MASH?

While pellets may be the best option in terms of ease of use and reducing wastage, some experienced poultry keepers think dry mash is a better form of feed. Because birds can consume their entire day's ration of pellets in half an hour, they soon become bored, especially in winter. By contrast, it takes hens about three hours of continuous feeding to eat the equivalent amount of mash. However, if you are keeping birds for exhibition, pellets are your only option as mash is too messy. Show birds, especially the bearded or muffed varieties, will end up with food-encrusted feathers around their beak if mash is used.

LEFT Dry mash takes longer to eat than pellets and may prevent your hens from becoming bored.

Intensively-reared chickens bought from commercial farms after their first season of egg-laying will almost certainly have been fed on mash. If you wish to change their diet, it is possible to wean them on to pellets and mixed grain. However, to build up their stamina when they first arrive, give them a warm mash mixed with 'treats' such as cooked vegetable peelings, cooked rice or crushed cornflakes. This will soon turn a poorly feathered, lacklustre hen into a blooming individual.

Where mash is being used, it is important to stir it well before feeding. Studies have found that mash ingredients such as ground limestone (included as a source of calcium) tend to sift through the food and settle at the bottom of the bag. Therefore, birds that are given food from the top of the bag may miss out on some of these vital components.

If for any reason you wish to change the type of food you give, make sure you do so gradually, over a period of seven to ten days. Any change-over should roughly take the following pattern: mix in a quarter volume of the new food to three-quarters of the old for two or three days; mix half and half for the same amount of time; give three-quarters of the new food mixed with a quarter of the old for another two or three days; then finally feed the new ration entirely.

ABOVE Split maize can sharpen up faded legs and boost the colour of egg yolks.

CEREAL FEEDS

It is pointless trying to give a combined feed of pellets and cereal at the same time, as most birds will eat the wheat first and leave the rest. As cereal is harder to digest, it is best given on its own as the afternoon or early evening feed, thus ensuring that the birds go to roost with a full crop. The afternoon feed is also one occasion where an exception can be made to the rule of feeding chickens indoors. If the corn is scattered around the outdoor run among dead leaves or other suitable scratching material, they will gain a great deal of amusement and exercise.

The best mix for a cereal feed is wheat and maize, but the ratio should not be more than around one-quarter maize to three-quarters wheat. That said, a little more maize can be fed during the winter months, as it provides the ideal nutrients to allow the bird to maintain body heat. Too much maize in a laying bird's diet is bad for its health as fatty deposits can build up around the ovaries, but a diet high in maize will fatten table birds more quickly and give their flesh that 'corn-fed' colour. Split maize is a very useful tool in enhancing the colour of yellow-legged breeds intended for showing.

Bags of ready-mixed grain can be bought, but you may find that they contain barley, which chickens and bantams dislike and often reject in favour of other grains. Therefore it is generally more effective, and less expensive, to buy wheat and mix in the maize yourself.

Sunflower seeds, linseed and other even smaller seeds intended for feeding to wild birds can also help keep domestic poultry amused. It is more economical to buy such items in bulk rather than in the smaller bags sometimes seen in pet shops and supermarkets, but this is only worthwhile if you use them up before they reach their expiry date.

BELOW Smaller bird seed can be given as a treat but should not form part of the daily cereal ration.

HOUSEHOLD SCRAPS

If you are looking for maximum egg production, household scraps should not form more than 20 per cent of your hens' daily food intake. If you are keen to utilize food scraps, you can boil them up and add them to mash, but make sure that the resultant mixture is moist but crumbly, not sloppy – if it breaks up easily when a mixing spoon is put through it, the texture is about right. Although this feed (called wet mash) is favoured by some poultry keepers, it is very time-consuming to prepare and you should put out only as much as the birds can eat in about half an hour or it will become sour and unpalatable.

Chickens enjoy leftover stale brown bread (white bread is as unhealthy for poultry as it is for humans) soaked in milk, cheese, rice, pasta and any other cereal-based scraps. Do not feed strong-tasting, mouldy or otherwise contaminated food, as this will, at the very least, taint the eggs and, at worst, cause digestive problems and even death. The discarded outer leaves of green vegetables can be given to chickens as part of their daily diet of greenstuffs.

GREENSTUFFS

Chickens need greenstuffs for the minerals and vitamins they contain, and in the case of free-range birds most of their requirements can be obtained naturally. Birds kept in runs or paddocks will also find greenstuffs in the form of grass or weeds, but those in grassless runs will have to be fed a supply. A few turfs of short grass can be thrown into the run at intervals, but it is far better to hang fresh leaves off the ground or to put them in racks daily. Whatever method is chosen, any leftovers must be removed each night and placed on the compost heap. In addition to greenstuffs, windfall apples and any sweetcorn that has gone starchy and is

ABOVE Grass is high in chlorophyll, which can improve the colour tone of yellow-legged chickens.

too hard for use in the kitchen will also be readily devoured.

If you have space in your garden, you may consider growing crops especially for feeding to your poultry. Useful feeds include lucerne, vetches, green cereals, rape, kale, millet and clover. The birds will obviously also appreciate kitchen vegetables that have gone to seed. Always try to give them the whole plant, as they will benefit from the grubs and minerals in the soil surrounding the roots as well as from the leaves and stems.

Weeds gathered in the countryside, provided they are free of agricultural or car-exhaust chemicals, can be another good addition to the diet. Young chicks in particular appreciate chopped dandelion leaves, plantains, shepherd's purse, watercress and, appropriately enough, chickweed.

GRIT

Flint grit is essential in allowing birds to digest food once it has entered the gizzard (see page 123). Because stones on the ground are usually rounded, they have little effect on grinding food into the necessary consistency, so flint grit is given separately. It can be purchased from any agricultural supplier – expect each bird to consume approximately 30g (1oz) per month. Rather than putting the grit in a separate trough, some poultry fanciers place a thin layer in the nest-boxes in place of hay or straw so that the hens pick up all their necessary requirements when laying.

Grit is also a means of supplying calcium, and although this is often included in manufactured feedstuffs, it can be given separately as oyster shell or ground limestone. Many chicken keepers believe that birds fed good-quality layers'

ration should not be given extra oyster shell because this will upset the calcium: phosphorus ratio and result in thin-shelled eggs, brittle bones and, in some cases, feather loss.

VITAMIN AND MINERAL SUPPLEMENTS

While vitamin and mineral supplements are available commercially, they are not normally needed because the manufactured pellets and mash always contain the usual daily requirements. Birds that have access to grass will usually find sufficient minerals and vitamins as they scratch around, although there may be certain times during the year – such as in winter, when rearing birds and at the moult – when supplements may prove beneficial.

LEFT Mixed grit is essential to the chicken's digestive system.

It is particularly important that sufficient vitamins are included in the diet of your breeding stock, otherwise the resulting chicks may hatch with curly, misshapen toes, splayed legs or twisted necks and breastbones. (It is important to note here that similar defects in incubator-hatched birds may be a result of bad incubator management and have nothing to do with genetic deficiencies.)

When you are showing birds or moving them long distances, there is always a possibility that they might become affected by stress. This is another situation where supplementation is useful, as a course of soluble vitamins obtained from your veterinary surgery and given in the drinking water for five to seven days, can help alleviate stress. As sachets of soluble vitamins have a long shelf life, they can usefully be kept on hand. Be sure to follow the vet's instructions and discard any medicated water that is more than 24 hours old.

WHEN AND HOW MUCH TO FEED

The amount of food required depends on the breed of your birds, their stage of development, the time of year and the methods and environment chosen to house them. Ideally, chickens prefer to feed shortly after descending from the roost and before laying their eggs. After some dusting and exploring, they may take a rest in the middle of the day, before feeding again in the mid- to late afternoon. If your feeding routine can reflect this, so much the better.

There are generally two options when it comes to feeding: ad lib and by hand.

As mentioned before, it is important not to overfeed your stock, as any surplus will encourage fat, unhealthy birds that will neither breed nor lay to the best of their ability, as well as the possibility of a plague of vermin. Most breeders prefer a programme of hand-feeding, as the correct quantities can be judged more accurately. However, where this is not possible (perhaps owing to work commitments), you may have to compromise, with pellets or mash fed in a hopper in the morning and an evening scattering of grain. Remember that chickens are daytime-loving creatures, so the last feed should not be given so late that they have insufficient time to fill their crop for the long night ahead before going in to roost.

When estimating how much to feed by hand, work on the basis that a fully-grown hen eats about 100g (4oz) of food a day, of which half (a good handful) should be grain. Particularly large breeds may require a little more – perhaps, 150g (6oz). For adult bantams, a 60:40 ratio of pellets to grain is considered by many breeders to be a better mix, but it is up to you to observe their daily food intake carefully in order to determine the correct quantity. If you find no trace of food an hour or so after the birds have been given their breakfast, increase the amount until the last particles are being picked up around lunchtime. Conversely, if there is food left from the morning when you take out the afternoon grain supplement, decrease the breakfast quantity accordingly. Very young chicks should be fed four or five times a day, but once they are reasonably active and strong, a small trough or hopper can be used for ad lib feeding.

WATER

Never let your birds go without clean, fresh water. How much they drink in a day depends on the time of year, the type of food being given and the size of the bird. As with feed, you must be the best judge of how much your flock drinks, something that can be estimated only by careful observation on your part.

WATER

Water is vital to chickens, helping to regulate their body temperature, digest their food and expel waste body matter. It is particularly important to a laying hen for the simple reason that a substantial portion of an egg consists of liquid and she needs to be able to produce this quota on top of what she needs for herself.

Never let your birds go without clean, fresh water. How much they drink a day depends on the time of year, the type of food being given and the size of the birds. As with feed, you must be the best judge of how much your flock drinks, something that can be estimated only by careful observation on your part.

However, as a rule of thumb, during normal weather conditions the average-sized bird needs to drink 300ml (just over ½pt) of water daily; in hot weather this amount can double or even treble. Although you may occasionally see your birds drinking from puddles, do not assume that they can be left to find their own resources – the bacteria found in that sort of water will almost certainly have a harmful effect on their health. For the same reason, it is best not to use rain water collected via guttering and drainpipes and stored in barrels. Suitable though this may be for the vegetable garden, you can never be sure exactly what impurities it contains. Likewise, be careful about using well water, which might contain damagingly high levels of sodium, chloride sulphates, magnesium and copper. Copper in particular tastes bitter and may cause liver damage to poultry. An abnormally high level of chlorine in tap water is also detrimental, as it will cause chickens to drink less. If you live in a hard-water area, you may find stains or residues left on the drinkers, but this will have no effect on the bird's health, laying performance or fertility.

Although healthy birds fed a correct

LEFT Drinkers are best placed outside the house to minimize wetting of the floor litter and prevent it being scratched into the water.
ABOVE Plastic feeders and drinkers are less durable than galvanized but easier to clean.

ABOVE Dustbins make inexpensive vermin-proof containers for chicken feed.

diet, given access to fresh grass and an abundance of clean water should need little in the way of additives, some breeders swear by adding a mix of apple cider vinegar and minced garlic to their birds' drinking water, maintaining that this has a beneficial effect on the birds' general health and immune system. If you want to try this, mix one or two cloves of garlic into 1l (1 ¾pt) of cider and add a cup or two of the resultant mixture to 4.5l (1 gallon) of water.

A teaspoon or two of glycerine added to water on a cold winter's morning will help to prevent it from freezing over during daylight hours. Also, when night temperatures fall, some ingenious chicken keepers put a 40-watt light bulb inside a clay pot in the hen-house, the resultant heat often being sufficient to prevent water from freezing overnight. However, great care must obviously be exercised here, and it is imperative that you seek the advice of an electrician before running the risk of setting the house on fire.

FOOD AND WATER CONTAINERS

When considering the size of feeder required, the important measurement is the outside edge. For a circular feeder, allow 2.5cm (1in) of the perimeter for each bird. For a trough, increase this length to 10cm (4in) per bird. For reasons of biosecurity (see page 160) it is best to feed your birds indoors, but if you need to feed them outside from a hopper, do not be tempted to buy one with a flat feeder pan. Feeder-pan bases must be conical so that the food flows into the trough freely and no unused or mouldy feed remains in the centre, something that would undoubtedly be the case with flat-pan models.

If you are feeding in the run or hen-house and are using troughs, these should be mounted on bricks or wall brackets so that they do not become contaminated with floor litter. For the same reason, circular feeders should be suspended from a cord or chain attached to a hook fixed to one of the roof beams.

Drinkers may be galvanized or plastic and come in several varieties. Allow 2.5cm (1in) of drinker perimeter per bird. Fountain drinkers are more hygienic on a day-today basis, but drinking troughs can be advantageous in the winter as the water they contain can be prevented from freezing over by floating glycerine on the surface each morning when you change it. For very small numbers of chickens or for bantams, earthenware dog bowls make good feeders and drinkers, as they are heavy enough to avoid overturning and are easy to keep clean.

Another option is to construct an automatic drinking system, which can be done quite easily by fitting a ballcock and valve into a heavy-duty plastic bowl. Water is supplied via an alkathene pipe fed from a mains water supply. It is important to check that the ballcock is working freely in order to prevent flooding, and to place the bowl on a level surface. Alternatively, low-pressure hanging drinkers can be obtained inexpensively from any agricultural stockist. Suspend them from a tripod made of stout stakes or hung from a crossbar attached between two posts. Water is supplied from a header tank, not directly from the mains, and the filter that prevents sludge from fouling the spring-loaded release washer must be cleaned weekly. Both types of automatic drinker can be kept clean simply by turning them over and tipping the contents into a bucket so that the ground does not become soggy; they will then automatically refill and be flushed through.

FOOD STORAGE

The shed where you keep your chicken feed should be cool, dry and well ventilated, and the feed must also be protected from dirt, dust and vermin. Provided that the shed cannot be accessed by family pets, it is a good idea to keep several baiting points continuously supplied with rat and mouse poison.

Do not store feed bags directly on a stone or concrete floor, no matter how dry it appears, as condensation and moisture will undoubtedly build up between the floor and bag. This is the reason why agricultural merchants store their food on pallets or shelving.

Do not use any manufactured rations that have gone past their expiry date. Not only is there a danger that such food will have gone off; any beneficial drugs or vitamins it contains will also be far less or even totally ineffective. If any food becomes damp or spoilt, discard it immediately.

If only relatively small quantities of food need to be stored, keep them in containers that come complete with lids – galvanized containers are preferable, as rats can chew through even the most substantial plastic. Second-hand metal food-storage bins can often be picked up at local farm sales and auctions. Some of these even have dividing partitions, enabling you to keep pellets and grain in the same container.

RIGHT On a hot day a chicken can drink approximately 300ml (10½fl oz) of water.

BREEDING
Chickens

AS YOUR ENTHUSIASM FOR KEEPING POULTRY INCREASES, THERE IS BOUND TO COME A TIME WHEN YOU WILL WANT TO REAR A FEW CHICKS, EITHER JUST FOR FUN OR TO REPLACE AGEING BIRDS. BREEDING FROM YOUR OWN STOCK BIRDS IS THE MOST SATISFYING OPTION, BUT IF YOU DON'T HAVE A COCKEREL YOU CAN BUY FERTILE EGGS INSTEAD AND HATCH THEM UNDER A BROODY HEN. BORROWING A COCK BIRD OF PROVEN PEDIGREE AND RUNNING IT WITH YOUR FLOCK OF HENS FOR A MONTH BEFORE RETURNING IT TO ITS OWNER IS ANOTHER ALTERNATIVE. IF YOU EXPLAIN YOUR PLANS TO YOUR NEIGHBOURS BEFOREHAND, MOST WILL BE PREPARED TO PUT UP WITH THE CROWING FOR SUCH A SHORT PERIOD OF TIME.

BREEDING PENS

Assuming that you own or can borrow a cock bird, he should be of the type described under Selecting a Breeding Cock on page 30. For their part, the hens must be mature and should conform to the breed standard if there is one. Body size is important, and it should go without saying that all prospective breeding stock must be healthy.

The cock-to-hen ratio will depend on the breed; some males are virtually monogamous, while others are capable of fertilizing the eggs of ten or more hens. However, on average one cock to six hens is about right, although you should seek the advice of an experienced breeder who has a good knowledge of your particular breed. If you want to produce birds for the table or for egg production, flock mating can be quite successful and involves running a number of males with selected groups of females at the rate of one cock to 10–15 hens, depending on the breed. Normally, however, single-male mating is practised (not to be confused with single mating; see box on page 141).

If all your birds have been bought from the same breeder, there is a good chance that they will be related. As long as they are not too closely related, there should be no problem, but repeated inbreeding can result in undesirable traits in your strain (see box on page 141).

Cock birds can be run with the flock year-round, although professional breeders sometimes house them separately until the breeding season, believing that this keeps the males keen and eager to do their job. If the cock is kept separate, introduce him to the hens at least a month before fertile eggs are needed for hatching. Eggs may be fertilized within a week of mating, but it is best to allow a little more time to be sure. If a male of another bloodline is subsequently introduced, it is advisable to allow a fortnight after the changeover before beginning to save any eggs for hatching.

THE FERTILE EGG

If you have bought in fertile eggs, leave them for 24 hours so that the yolks can settle before incubation (see page 150). They do not need to be kept warm at this stage, as development will simply cease if the surrounding temperature is below 20°C (68°F). If the egg is then brought back up to a suitable temperature within a reasonable time, either under a broody hen or in an incubator, the arrested development will resume and proceed as normal.

The length of time a fertile egg remains viable depends on the conditions in which it is stored. For the best chances of hatchability, collect eggs as quickly as possible and store immediately at a cool temperature, as they will keep longer than those cooled more slowly through being left in the nest-boxes. The ideal storage temperature is 12.7°C (55°F). The storage time also depends on whether the egg will be hatched artificially or naturally with a broody hen. Generally, eggs hatched under a broody can be kept for around ten days, whereas those going into an incubator should not be more than a week old. It is interesting to note that chicks from eggs that have been stored are usually smaller than those from eggs that have not, and they also tend to hatch several hours later.

RIGHT Generally, the calm, heavier breeds make the best broodies and mothers.

BROODY HENS

You might notice that one of your hens stays in the nest-box longer than usual. If she remains there for a couple of days, put your hand under her. If she tries to peck at it, ruffles her feathers and squats further down into the nest, she is broody and ready to hatch some eggs. However, before you give her fertile eggs to hatch, check that she is of a suitable type. Most dual-purpose fowls sit well, as do cross-bred stock from heavy breeds.

The best location for a sitting box is in a shed well away from other birds. This should be well ventilated, fairly dark and free from rats – an old-fashioned coop and run is ideal. Cut a grass turf the same size as the box and place it upside-down in the bottom. This prevents eggs from rolling out when the rest of the nest is built up with hay, and it also helps to retain moisture and humidity, both of which are important for incubation. Do not move a broody hen from the hen-house to the sitting box in daylight; instead, wait until dusk. Make sure she is free of mites and lice (see pages 162–3), giving both her and the nest a good dust with flea powder before she settles down.

Rather than risking fertile eggs straight away, it is a good idea to put a dummy egg

RIGHT A broody hen should be kept apart from the remainder of the flock.

in the nest and then give the hen a day to acclimatize to her new surroundings. If, after 24 hours, she is sitting happily and tightly, encourage her out to feed and drink. You can then replace the dummy egg with a clutch of real ones. Just how many eggs to incubate will depend on the size of the hen, but as mentioned earlier, it is often thought that an odd number should be set, as this makes it easier for her to turn them.

The broody should leave the nest once a day to feed and empty her bowels. She may be so intent on carrying out her matronly duties that you have to lift her off – if this is the case, make sure you feel around her when you pick her up in case any eggs are tucked in her feathers. Keep an eye on her and do not let her stay away from the nest for more than 20 minutes.

INCUBATORS

Hatching eggs in an incubator can be quite complicated and the results are nowhere near as successful as those you will achieve using a broody hen. However, there are many small electric incubators on the market that may be worth experimenting with. If you are intending to rear poultry as a commercial sideline at some future date, an incubator will be essential – you cannot guarantee that broody hens will be around when you want them. Each incubator model is different, so it is important that you read and understand the manufacturer's instructions if you are to have any hope of success. If you want to hatch just a few eggs, a member of your local poultry club may be prepared to incubate them for you alongside their own.

Although there are many incubator models, there are only two basic types: cabinet machines and those that work on a still-air principle. The latter are used widely and very successfully, have few working parts that can go wrong, are easy to

understand, and are inexpensive and simple to maintain. Most of the modern models are electric, although you may still come across an older oil- or paraffin-heated version for sale at a farm auction. However, do ask someone experienced to check it over for you and make sure that all the essential working parts are present before you make your purchase.

Cabinet machines usually have more than one level of setting trays and differ from still-air models in that a fan circulates the air constantly to ensure an even temperature. Because of their size and capacity, they are normally used only by commercial chick producers. A separate hatcher is also required with a cabinet machine, to which the eggs are transferred a couple of days before they are due to emerge from the shell.

Humidity is an important aspect of incubator management, as an egg loses water through evaporation from the moment it is laid. During incubation, this rate of evaporation must be controlled by adding water according to the manufacturer's instructions. Too great a loss of water can cause the embryo to die, while too much can drown it. Ventilation in the incubator is also crucial and serves two purposes: supplying oxygen and removing harmful gases such as carbon dioxide. Some machines have an arrangement of two or three felts, which are removed one by one at intervals of a week and let out the stale air.

Most modern incubators have fully-automated turning devices, although you must check daily that the turner is working by noting the position of the trays or eggs. In older models, the eggs must be turned by hand at least twice a day to prevent the developing embryo from sticking to the side of the shell.

COMMON BREEDING METHODS

Line breeding uses two unrelated birds, the offspring of which are mated together in a systematic way to produce two distinct lines of chicken based on each of the parents. The lineage of any offspring, no matter what generation, is traceable to one or other of the original parent birds and, in theory, the gene pool of the first generation can be re-created endlessly.

Inbreeding is a method used to fix desirable genes, although it is not recommended for the novice as undesirable genes may also be passed on. The system is similar to line breeding, but brother–sister mating is used.

Outcrossing is when a totally unrelated bird is brought in to add new blood to an existing strain in situations where that strain has become weak through repeated line breeding or inbreeding. Like inbreeding, outcrossing can be risky, as unwanted traits may be introduced if the new bird has not been selected with the utmost care.

Single mating involves mating a single male to a single female and is uncommon except in the practically monogamous Asiatic breeds such as the Malay. It may also be the first step in a line-breeding strategy or as an outcross. (Double mating has a totally different meaning and is explained on page 31).

ABOVE Candling is shown here in a commercial situation, but the basic principles are the same for the small-scale poultry breeder.

OPPOSITE Chicks will quickly become curious about the outside world but will need to stay close to 'mum' for about three weeks to keep warm.

CANDLING

When an incubator is being used, it is advisable to check the eggs each week to ensure that the embryos within them are developing correctly. This is best achieved by candling, which involves holding each egg below eye level in front of a bright light, preferably in a darkened room.

Candling lamps can be bought or made inexpensively and consist of an electric light bulb placed inside a box, on top of which is an egg-shaped hole. The egg is held over the hole, so that no light escapes around its sides (most candlers have a rubber seal).

It is recommended that candling is carried out weekly to eliminate infertile, dead or damaged eggs. It is also a useful method for checking humidity by monitoring the air sac. At the first candling, at five to seven days, a fertile egg is indicated by the presence of the blood vessels of the embryo (these

resemble a multi-legged spider), and the air sac should be evident at the broad end of the egg. As incubation progresses, the air sac becomes larger until, by hatching time, it takes up nearly a quarter of the egg, the remainder of which will appear almost completely blacked out. If the air sac is deemed to be too large at any stage, the humidity level inside the incubator must be raised. If the air sac is too small, lower the humidity, either by increasing ventilation or by reducing the surface area of water trays in the incubator.

At 14 days, infertile eggs will appear clear and should be removed from the incubator to prevent possible contamination. This also improves the chances of the remainder hatching.

Candling and the process of incubation is too complicated to discuss in depth in a book of this nature, and it is therefore recommended that anyone interested in the subject reads the comprehensive, revealing and informative *New Incubation Book* by Dr A.F. Anderson Brown and G.E.S. Robbins (Blaine, WA: Hancock House Publishers, 2002).

RAISING CHICKS

Once the eggs begin to hatch, confine the broody hen to the nest-box and leave well alone. As soon as the hatch has finished and the chicks are dry, transfer them to a coop along with the hen. If the coop is in a shed, make sure that the hen cannot take her chicks too far from the nest by building a temporary pen. The sides of this must be tall enough to prevent her from flying over, otherwise the chicks will be left behind and may become chilled. If the chicks have been hatched in an outside coop, it should now have a run attached to it and be moved on to fresh, short grass. Feed the chicks on proprietary chick crumbs for the first three weeks and make sure their water container is shallow enough that they cannot drown.

If the chicks have been hatched in an incubator, it is possible to persuade a broody hen to become a foster mother to them. The hen must have been sitting on a clutch of dummy or even infertile eggs for at least a week and be well settled into the nest. Plan the arrival of the chicks for late evening and make sure that they can be kept warm for a couple of hours – buy a chick box for the purpose, or make one by placing a cloth-covered hot-water bottle in the bottom of a cardboard box. As dusk approaches, very gently take away a couple of the eggs under the hen and replace them with two or three chicks. The hen will not like being disturbed, and she will fluff up her feathers and may even try to peck the back of your hand as you slide the chicks underneath her. Tuck them in among the remaining eggs, back off slowly

Once the chicks are three weeks old, the heat source can be turned off in the day during spells of warm weather. If only small numbers of chicks are being reared and the weather is hot, you could take the birds to a small wire pen on the grass outside for a few hours. However, make sure that the pen is cat-proof, place it in the shade and take the chicks inside immediately should it start to rain.

FEEDING ARTIFICIALLY REARED CHICKS

Feeding and drinking vessels in artificial brooders should initially be placed around the heating area so that the chicks do not have to go far to look for them. Small jam jar-type drinkers are ideal, although commercial plastic chick drinkers are available. Chick feeders can also be bought, but almost any shallow pan will do provided that the birds aren't able to scratch their food all over the place.

Do not worry if you don't see the chicks eating during the first 24 hours as they still have enough nutritional reserves from the egg yolk. The inclusion of some finely chopped hard-boiled egg, sprinkled over the chick crumbs, will help to arouse their curiosity and also provides extra protein for the first couple of days. Do not mix up too much at any one time – as with wet mash for older birds, it will soon go sour.

and leave her to settle again. After an hour or so, if all is well, take away the remaining eggs and replace them with the other chicks. For maximum success the process must be carried out slowly and quietly.

USING AN ARTIFICIAL BROODER

If you do not have a broody hen to rear your incubator-hatched chicks, you will need an artificial brooder. This is usually located in the sectioned-off corner of a shed, with warmth provided by gas heaters or infrared lamps. If you have only half a dozen or so chicks, a conventional light bulb fixed in an upside-down terracotta plant pot will give sufficient heat and the chicks will cluster around it happily.

Whatever heating system you choose, the room, shed or building must be maintained at a constant temperature, so forward planning is required. A metal roof, a sunny day and the heat from a lamp, for example, would be a bad combination. Even with suitable surroundings, always keep a check on the ambient temperature inside the brooder, especially around midday.

Chicks need to learn where the source of heat is straight away. If rearing is carried out in a big shed, this is best achieved by making a circular surround of cardboard or hardboard and keeping the chicks within it. Suspend the heat lamp from the ceiling at a height above the floor that ensures the temperature is around 32°C (90°F). Increase the lamp's height progressively each week until, by the third week, the temperature at floor level is down to 21°C (70°F). You can usually tell if you have got the temperature right by watching the chicks. If they are huddled under the heater, they are too cold; if they are pressing against the surround, they are too hot and either the ring should be enlarged or the heat lamp raised. After a few days, the cardboard surround can be increased to give the birds more space, eventually being removed altogether.

ABOVE LEFT These chicks have a mother to show them what to do – if you don't have a suitable hen, you may have to give 'pecking lessons' yourself.

RIGHT A Silkie or Silkie-cross bantam is considered by many to make the best broody.

All About
EGGS

'WHAT CAME FIRST, THE CHICKEN OR THE EGG?' THIS CONUNDRUM HAS PUZZLED HUMANKIND DOWN THE AGES. WAS IT THE HEN WHICH AFTER ALL LAYS THE EGG, OR WAS IT THE EGG, FROM WHICH THE HEN HATCHES?

IN THE BEGINNING…

The Bible states that – along with other birds – the chicken was created, not the egg, but according to the theory of evolution, all new species develop from mutations. If the new characteristics generated through mutation are successful for survival, they are passed on to successive generations and a new species is formed. Genetic material does not change during an animal's lifetime, so the DNA of an embryo inside an egg is the same as that of the chicken it hatches into. The first chicken was a mutation of its avian parents, but its life began inside the egg, so it was the egg that came first – the remarkable, wholesome, miraculous egg.

And what an egg it is! Not only are chickens' eggs important emulsifiers used in all types of cooking, they are one of the most nutritious foods money can buy. Even though they each contain just 307kJ (75 calories), eggs provide all the essential amino acids and minerals required by the human body, and are an important source of vitamins A, B and D. They also supply a complete protein that is of a higher quality than any other food protein.

Hens fed a special diet containing 10–20 per cent flaxseed produce eggs enriched by the polyunsaturated essential fatty acid omega-3. This naturally occurring substance, commonly found in fish and fish oils, helps to lower blood triglyceride levels, promotes good vision and is required for normal growth and development in the human body.

Some people avoid eating eggs in the belief that the high cholesterol levels found in the yolk are harmful. It is thought, however, that egg yolk may actually lower total body levels of low-density lipoprotein, the undesirable form of cholesterol, while raising levels of high-density lipoprotein, or 'good cholesterol'.

EGG FORMATION

An average egg weighs 60g (2oz) and consists of three parts: the shell, which is made of calcite (a crystalline form of calcium carbonate); the albumen, or white, which is made up of water and protein; and the yolk, which is the most nutritionally valuable part of the egg.

Eggs develop individually in the ovary, then detach and slip into the oviduct, a long tube that ends in the vent, or cloaca, through which they exit the body (droppings are passed through the same orifice). As an egg travels through the oviduct, it rotates continually within the tube, the movement twists structural fibres called chalazae into rope-like strands that anchor the yolk in the white from opposite ends of the egg. Once a fertile egg has been laid, the chalazae keep the germinal disc of the yolk on the uppermost surface, near the heat of incubation. This is why fertile eggs that have been transported must be given 24 hours to settle before incubation is attempted.

The hen's oviduct has two parts: in the first, the white forms around the yolk; and in the second, the shell is made and the pigment of the shell is deposited. Each egg takes 3–5 hours to pass through the first part of the oviduct and a total of 15–20 hours before it is ready to be laid.

While the colour of the shell has no relevance to the nutritional value of the egg, it is related to the ear or cheek colour of the hen (see page 12). Chickens' eggs range from snow-white to dark brown, covering all shades of beige in between, and some may be speckled and even blue or green (as produced by the South American Araucana breed).

The colour of the yolk is directly associated with the hen's diet. A very pale yolk can indicate that the hen lives in overcrowded quarters, is underfed or lacks

EGGS & SALMONELLA

The *Salmonella enteritidis* bacterium, which causes salmonella food poisoning, is found in the faeces of many animals, including chickens. Because hens sit on their eggs, there is a risk that the bacterium can enter the eggs through their porous shells. (Interestingly, it has been found that Maran eggs remain free from salmonella infection, possibly because the pores in their eggs are small and the bacteria cannot penetrate the shell.) Stringent cleaning and inspection procedures put in place since the 1970s have reduced the incidence of external egg contamination to virtually nil in commercial flocks, although *Salmonella enteritidis* can still enter eggs in the ovaries of hens before the shells have formed.

LEFT While commercially available eggs are almost invariably brown, keeping your own chickens means you can have the pleasure of collecting new-laid eggs in beautiful shades of chalky white, blue-green or olive.

EGG GAMES

Egg jarping, or egg tapping, is a traditional Easter game from the north of England and is rather like conkers. Players tap each other's uncooked eggs in turn until one breaks. The victor then goes on to the next round, until eventually there is only one good egg left – the winner.

Egg rolling takes place on Easter Monday, and again is particularly popular in the north of England. Hard-boiled eggs are rolled down a hill and, depending on the variation played, the winner is the person whose egg rolls the furthest, survives the most rolls or lands nearest a target. Easter egg rolls are also popular in the USA – the event on the lawn of the White House is the best known and has become an annual tradition. The first official White House Easter Egg Roll was held in 1878. Today, the President and First Lady invite children from across the country to attend the celebration.

greenstuff, whereas a bird fed a diet rich in xanthophylls (the yellow pigment from the carotenoid group found in green leaves) will produce a darker-yellow yolk.

Occasionally, a hen may lay an egg with no yolk at all or one with a double yolk. These are both the result of unsynchronized production cycles, and usually occur at the start or end of a laying period. The so-called 'meat spot' in the egg is, in fact, a small deposit of blood caused by the rupture of a blood vessel during its formation. It does not indicate a fertile egg. It is not harmful, but can easily be removed with the tip of a knife.

A cloudy white is a sign of freshness and is caused by carbon dioxide – this gas is present naturally in the white when the egg is laid, and decreases over time as it escapes through the porous shell.

EMBRYO DEVELOPMENT

All birds will lay a clutch of eggs before starting to incubate them, and a hen is no different. She will continue to lay as long as her eggs are removed daily and she thinks she hasn't produced enough for a clutch. When enough eggs have been laid, she will go broody and lose her breast feathers in order to warm the eggs and begin incubation. Her behaviour also changes: she will remain on the eggs, fluffing up her feathers and making croaking noises if approached, and will leave the nest only once a day to eat, drink and defecate.

The embryo of a large fowl normally develops inside the egg for 21 days, whereas that of a bantam has a slightly shorter incubation period of 19–21 days. As it grows, its primary food source is the yolk. It is important that the embryo doesn't become stuck on one side of the egg, so the broody hen will turn all her eggs several times a day. After 21 days, the chick pecks its way out of the shell with a horny growth on its beak known as the egg-tooth.

Once the chicks have hatched, the broody hen will remain on the nest for a further 24–48 hours. After this time, any eggs that have not hatched will be left behind when she takes her chicks on their first outing.

Certain breeds of chicken, in particular the Mediterranean varieties such as Leghorns and Minorcas, have had their broodiness bred out of them so that they produce more eggs. It is also rare for hybrids to go broody. As you may not always want to be hatching chicks, there are various methods of dissuading a hen from being broody. Sometimes, just isolating her from the nest-box will do the trick. A more stubborn hen can be put in a cage with a wire bottom, so that the airflow cools her underside and discourages her from sitting. Alternatively you could try giving her a 'clutch' of ice cubes, although it may take several attempts before this cure proves effective.

RIGHT A clutch must be complete before incubation starts, so each day the hen will add an egg and then leave it to cool.

EGG STORAGE

Always store eggs with the point down so that the yolk stays centred. As the shell is porous and will absorb smells from its surroundings, eggs should also be kept away from strong-smelling foods such as fish or scented products like soap. The porosity also makes it unwise to wash eggs. If they are very dirty, brush or wipe them instead.

Many people keep their eggs in the refrigerator, although they are better off in a cool place but not chilled. Refrigerated eggs frequently break when boiled, the absorption of bacteria through the porous shell is increased in the enclosed space, and the lower temperature can cause the protein to break down, reducing nutritional benefits. Take your cue from supermarkets and other food shops, which always store eggs on unrefrigerated shelves.

Raw eggs can be frozen very successfully and can be kept for up to a year. If frozen whole, the shells will crack, but the eggs can still be used for baking if defrosted carefully. A more successful way of freezing eggs is to put spare whites or yolks into ice-cube trays, defrosting the quantity you need as necessary. Cooked eggs do not freeze well, as they tend to be tough and rubbery when defrosted.

To tell if an uncracked egg is fresh, drop it into a bowl of water. If it sinks, it is fresh, but if it floats it is bad and bacteria have entered through the porous shell and have created gas inside. To tell if an uncracked egg is raw or hard-boiled, spin it, then abruptly stop it from spinning and let go. If it starts to spin again, it is raw as the liquid inside will continue to rotate. If it remains still, it is hard-boiled.

RULES AND REGULATIONS

If you keep large numbers of birds or plan to sell your eggs anywhere other than from your home, there are a number of regulations you must abide by. Some of these rules relate to food safety and egg traceability, while others are concerned with bird health.

In the UK, if you own more than 350 hens you must be registered with the Egg Marketing Inspectorate (this is free) and your eggs must bear a legible date stamp, applied with food-grade ink, and your producer registration number to allow traceability. The eggs must also be graded by size, which can be done only at a registered packing station. If you own fewer than 350 hens but wish to sell your eggs, they must also be stamped with a producer registration number and you can only sell them ungraded. You do not have to mark eggs sold at your farm gate or delivered by hand to your customers, but they cannot be used in the catering industry.

Producer registration numbers are used throughout the European Union and look like this: 1UK 56789. The first number refers to the way the hens are reared: 1 = free-range; 2 = barn-reared; and 3 = caged. The letters refer to the country of origin, while the final five digits are unique to the producer.

If you keep more than 50 birds of any breed or species commercially in the UK, you must by law register them on the Great Britain Poultry Register. This is part of the programme set up by the Department for Environment, Food and Rural Affairs (Defra) to improve risk assessment and monitoring of bird-flu outbreaks. Bird flu does not pose a danger to egg consumers, as the virus cannot survive cooking. For your eggs to be called organic, the layers must have

EGG-EATING

Egg-eating is a habit chickens often acquire through curiosity. It starts when a thin- or soft-shelled egg is laid or a normal egg is broken and one of the birds in the flock pecks at it and decides it likes the contents. The bird will then do the same to other eggs it may find.

Once egg-eating has become a habit, it can be very difficult to stop. Clean, dark nest-boxes will keep eggs out of sight and therefore reduce temptation. In addition, blown eggs filled with strong mustard can be put in the nest-boxes and may taste unpleasant enough to cure a bird, although you may have to do this for several days before succeeding.

BELOW Painting eggs is a popular activity for children at Easter.

EGG SIZES

Egg sizes in the UK now follow the European convention:

Small (S): less than 53g (1.9oz)

Medium (M): 53–63g (1.9–2.2oz)

Large (L): 63–73g (2.2–2.5oz)

Very large (XL): greater than 73g (2.5oz)

access to an outside area all year round or be fed sprouted grains for any period they are kept indoors. All feed must be certified organic, cannot include meat by-products, and no antibiotics can be given. In the UK, eggs denoted as class A are the highest quality and may not be cracked, should have a normal shell and should not be washed – it is considered preferable to produce a clean, quality egg in the first place, as this indicates high standards of production. Class B eggs are of a lower quality and may be cracked or dirty.

EGGS IN FOLKLORE

There are many superstitions surrounding eggs. For instance, it was believed that the tenth egg laid in a batch would always be the largest, but to find a small egg was bad luck. Even unluckier was an egg with no yolk, as this was said to have been laid by a cockerel. Sailors were advised never to mention the word 'egg' if they were to avoid misfortune at sea. And anyone eating a boiled egg should poke a hole through the bottom of the shell when they have finished to prevent a witch from using it as a boat, but shouldn't throw the shell on a fire or the hen will never lay again and a storm may brew at sea.

It was said that if a girl wants to find out who she will marry, she should boil an egg and then fast for a day. She should then take out the hard-boiled yolk and fill the hollow in the white with salt. While reciting an incantation to St Agnes, the girl must eat the salty egg, including the shell. The next man she sees will be the one she will marry, although if she takes a drink before sunrise, no matter how thirsty she becomes, her future husband's identity will not be revealed.

LEFT Whatever their colour, eggs are one of the most nutritious foods money can buy.

EGGS AROUND THE WORLD

Thousand-year-old eggs are a highly prized Chinese delicacy, but despite their name they are, only around ten weeks old. The eggs are preserved in a coating of clay, ash, lime and salt, during which time the chemicals in the clay soak through the shell, turning the egg a translucent blue or green colour and producing a slightly fishy taste.

In Imperial Russia, giving eggs was not restricted to Easter. The practice was so popular that the Romanov royal family employed jeweller Carl Fabergé to create wonderfully ornate gem-encrusted eggs, to be given as presents at any celebration.

The word 'cockney' is thought to have derived from 'cock's egg', the name for a small, malformed egg occasionally laid by a young hen. The term was applied by country folk to townsfolk generally because of their reputed ignorance of country life and customs (cocks don't lay eggs). In the 17th century specifically it came to denote someone born within the sound of Bow bells – the bells of the Church of St Mary le Bow in the City of London.

The first-century AD Roman historian Pliny the Elder wrote about a druid's egg, produced by the joint labour of several serpents and buoyed in the air by their hissing. Whoever possessed such an egg was sure to prevail in every contest and be courted by those in power.

'Columbus' egg' is the term used to mean a task that is easy once you know the trick. The story is that in reply to a suggestion that other explorers might have discovered America had he not done so, Christopher Columbus is said to have challenged the guests at a banquet given in his honour to make an egg stand on end. When none succeeded, he flattened one end of his egg by tapping it against the table and so stood it up, thus indicating that others might follow but that he had discovered the way.

ABOVE Commercially available eggs are almost invariably brown, but your own chickens may produce chalk-white, blue-green or even olive-coloured eggs.

USING UP YOUR EGGS

Most chicken keepers will at some time have an excess of eggs, usually in the summer months, and especially when the flock are fairly young. However many you give away to friends and neighbours, there may well be days when you are looking for inspiration on interesting ways to eat eggs.

Whole eggs

❍ Hard boiled eggs can be curried, stuffed, turned into Scotch eggs, added to kedgeree or salade niçoise.

❍ Poached eggs are delicious by themselves on toast, or you can add cheese sauce to create egg mornay, or sit them on spinach for oeuf Florentine.

❍ Oeuf en cocotte is simply an egg popped in a ramekin with a little cream and butter on top and baked in the oven.

❍ Omelettes come in all shapes and sizes – almost anything can be added but useful standbys are cheese, mushroom, tomato and ham. A variation is frittata or tortilla, which is basically an omelette that is not folded or turned. For a bit more bulk, add diced, cooked potato and onion.

❍ For something really special try scrambled egg and smoked salmon.

❍ Don't forget that old favourite, french toast (eggy bread) – a slice of bread soaked into beaten egg and fried till crisp and golden.

❍ Quiches, flans and soufflés – the sky's the limit here...

❍ A simple family pudding for using up eggs (and milk) is crème caramel.

Egg whites

Apart from meringues, egg whites are the basis of soufflés, mousses and sorbets. Apple snow is simply stewed apple with whisked egg white folded in.

Egg yolks

Zabaglione, mayonnaise (add garlic and you have aioli), custard, pancakes, crème brûlée. Some chocolate cakes and torte use frightening amounts of egg yolks – now would be the time to make one.

RECIPES

CURRIED EGGS

Ingredients

4 eggs, hard boiled
1 tbsp butter
1 onion, chopped finely
300ml (½ pint) of water
2 tsp curry powder
2 tbsp cornflour
1 tbsp sugar
1 tbsp vinegar
Seasoning to taste

Preparation

1. Brown the onions in the butter.
2. Add the curry powder, seasoning and sugar to the onions and blend well.
3. Mix the cornflour, a dash of the water and the vinegar, stirring to remove all lumps. Add to the onion mixture, stirring well.
4. Add the rest of the water to the mixture and bring the mixture to boil over a low heat, stirring all the time.
5. Let the curry sauce simmer for 10 minutes, stirring occasionally.
6. Cut the hard-boiled eggs lengthways and add to the curry mixture. Remove from heat and leave the eggs in the sauce for 5 minutes before serving. Serve on buttered bread or rice.

DIJON-DEVILED EGGS

Ingredients

6 hard-boiled eggs
2 tbsp mayonnaise
1 ½ tsp Dijon mustard
1 green onion, thinly sliced (slice a little of the green and keep separate from white)
A few leaves of chopped fresh flat-leaf parsley
Freshly ground black pepper
Salt, to taste
Paprika

Preparation

1. Halve or quarter the eggs and scoop the yolks into a small bowl. Mash well.
2. Add mayonnaise and Dijon until desired consistency is reached.
3. Stir in the white part of sliced onion and most of the chopped parsley. Add salt and pepper to taste.
4. Using a small teaspoon or pastry bag, fill egg white halves or quarters. Sprinkle with sliced green onion and the remaining parsley. Sprinkle with a little pepper and/or paprika.

Makes 12 halves, 24 quarters. Recipe can be doubled.

EGGS FLORENTINE

Ingredients

800g (28oz) spinach, lightly steamed, (measure after cooking)
12 eggs
Salt and pepper to taste
½ lb of cheddar
300ml (½ pint) white sauce
300ml (½ pint) buttered crumbs

Preparation

1. Grate cheese and melt into white sauce to make a cheese sauce. Put warm spinach in a greased ovenproof dish.
2. Make indentations in the spinach and break one egg in each depression. Season with salt and pepper.
3. Pour cheese sauce over all. Sprinkle with crumbs and bake at 180°C (350°F) until brown, usually about 25 minutes.
4. Serve it with a crusty french bread and buttered baby carrots.

Serves 6

HOLLANDAISE SAUCE

Ingredients

6 tbsp white wine vinegar
A few black peppercorns
1 bay leaf
225g (8oz) butter
4 egg yolks

Preparation

1. Put the vinegar, peppercorns and bay leaf into a saucepan and boil until reduced to 1 tbsp - watch carefully, as this happens very quickly.
2. In another pan, heat the butter until it melts and bring to the boil.
3. Put the egg yolks into a food processor and, with the machine running, strain in the vinegar and then very slowly pour in the boiling butter. The sauce is ready when it is thick enough to coat the back of a spoon and should be lukewarm rather than hot.

Variation To make paloise sauce, start off with the basic hollandaise sauce and add 1 tbsp mint sauce and 1 tbsp chopped fresh mint. This accompaniment turns roast lamb into something really special.

Preparation time: 4 minutes
Cooking time: 2 minutes
Makes about 275ml (½ pint)

LEMON CURD

Ingredients

225g (8oz) butter, at room temperature
450g (1lb) caster sugar
5 eggs
Juice and finely grated rind of 3 or 4 lemons

Preparation

1. Cream together the butter and sugar in a bowl, then beat in the eggs one at a time. Slowly add the lemon juice and grated rind. Don't worry at this stage if it looks as if it has curdled – it hasn't!
2. Place the bowl over a saucepan of boiling water and, using a balloon whisk, gently whisk every now and then until the curd looks shiny and opaque, has stopped foaming and coats the back of a spoon – this may take anywhere from 10 to 20 minutes. Be careful not to let it boil as this may well cause it to curdle.
3. Pour into sterilized jars.

Preparation and cooking time: 15-25 minutes
Makes 3–4 jars

MAYONNAISE

Ingredients

2 egg yolks
¾ tsp salt
½ tsp powdered mustard
⅛ tsp sugar
Pinch cayenne pepper
4 tsp lemon juice or white vinegar
Olive or other salad oil
4 tsp hot water

Preparation

1. Beat yolks, salt, mustard, sugar, pepper, and 1 tsp lemon juice in a small bowl until very thick and pale yellow. (If using an electric mixer, beat at a medium speed.)
2. Add about ¼ cup of oil, drop by drop, beating vigorously all the while. Beat in 1 tsp each lemon juice and hot water. Add another ¼ of a cup oil, a few drops at a time, beating vigorously. Beat in another tsp each of lemon juice and water. Add ½ cup oil in a very fine steady stream, beating constantly, then mix in remaining lemon juice and water; slowly beat in remaining oil.
3. Cover and refrigerate until needed. Do not keep longer than 1 week.

PICKLED EGGS

Ingredients

For the spiced vinegar:

1.1 litres (2 pints) malt, wine or cider vinegar
25g (1oz) mixed pickling spice; or ½ tsp ground cloves
½ tsp mace
½ tsp allspice
½ tsp ground cinnamon
½ tsp ground ginger (or a slice or two of fresh root ginger)
a few peppercorns
12 eggs

Preparation

1. To make the pickling vinegar, simply combine all the ingredients in a bottle and keep for 6–7 weeks before using, giving it a shake occasionally. If you can't wait that long, bring the vinegar up to the boil with the spices, leave to cool and use immediately.
2. Hard-boil the eggs for 8–10 minutes, giving them an occasional stir to try to keep the yolk in the middle as they cook. Plunge them into cold water and, when cool, shell them and pack them into clean, dry jars. Pour over the vinegar to cover, seal and store in a cool, dark place for at least 2 weeks before eating.

Preparation time: 2 minutes Cooking time: 8–10 minutes
Makes 1 jar of 12 eggs or
2 jars of 6 eggs

HEALTH
& Care

GOOD HYGIENE IS ESSENTIAL TO A BIRD'S HEALTH, PRODUCTIVITY AND WELFARE. IT IS EASILY ACHIEVED BY KEEPING ACCOMMODATION CLEAN, GIVING THE BIRDS PLENTY OF ROOM AND FEEDING THEM A BALANCED DIET. HOWEVER, SOMETIMES DISEASE CAN OCCUR IN EVEN THE HEALTHIEST OF CHICKEN YARDS, USUALLY BROUGHT IN ON THE WIND, VIA THE DROPPINGS OF WILD BIRDS, OR EVEN ON THE FEET OF HUMAN VISITORS. THEREFORE, MAINTAINING BIOSECURITY IS AN IMPORTANT FACTOR IN ENSURING THE HEALTH OF YOUR STOCK.

WARNING SIGNS

Most chickens like to be part of the flock, so it is usually an indicator that all is not well if a bird is moping in a corner on its own. The birds should always be busy scratching, dusting and feeding. Their combs should, in most breeds, be red and waxy and their eyes always bright. Drooping wings, ruffled feathers, a loss (or sudden gain) of appetite and loose droppings stuck to the feathers around the vent area should all be treated with suspicion.

If the birds are tame enough, occasionally pick them up to check under their wings and in other downy areas for signs of fleas, lice and mites. At the same time, feel each bird's breast for any sudden loss of weight.

MOULTING

Most birds shed their feathers annually in the late summer or early autumn, although young birds moult twice during their first six months of life. A partial moult also sometimes occurs in the early part of the year. This is often limited to the neck, especially in the case of point-of-lay pullets that have been given a layer's ration too early on in life.

If correctly cared for, a healthy young chicken will take around six weeks to change its feathers, whereas the process in older birds can take double this time. Some of the lighter breeds will also take less time than heavier, dual-purpose breeds. The large wing and tail feathers of all birds are replaced slowly in a specific sequence over an extended period of time. This sequence has evolved over thousands of years to ensure that birds do not lose too many feathers at once and so are always able to fly and escape from danger.

It is better for the bird if the moult takes place quickly and for this to happen it must be in good condition. The later in the season a moult starts, the longer the process will take. Because egg-laying decreases or even ceases altogether in moulting birds, this means that a laying bird will remain unproductive for longer.

To get egg-laying breeds back into production as soon as possible, an early moult can be artificially induced by moving the birds into a different environment or by radically altering their diet. As soon as most of the hens in the flock are moulting, you can restore their usual feeding routine and give them an extra boost of linseed meal, soybean meal and cod-liver oil. Once they have produced their new feathers, increase the proportion of animal protein in their diet to encourage laying.

WORMING

Some changes to legislation made in late 2009 might cause chicken keepers problems regarding the way in which they worm their birds. The Veterinary Medicines Act governs how veterinary medicines are authorized, manufactured, supplied and used. The Veterinary Medicines Directorate (VMD) is the regulatory authority that looks at responsible, safe and effective use of veterinary medicines. It defines a medicine as 'a substance or combination of substances presented as having properties for treating or preventing disease in animals'. All products that make a claim to do this must be licensed by the VMD and those that are can be identified by a 'Vm' or marketing authorization number on the packet. Flubenvet wormer, available in a 60g (2oz) pack is, however, exempt from the ruling and may be used legally by poultry hobbyists. One pack will treat about 20 birds and is available from most vets, pharmacists and qualified animal-health advisers.

BIOSECURITY

Usage of the term 'biosecurity' has become more common since concerns over avian influenza (see page 164) have come to the fore. It is, however, really just a buzz-word for common-sense health measures. By periodically removing droppings from the run, feeding chickens in an area that is not accessible to wild birds and placing a disinfectant foot-bath at the gate to your poultry yard, you are actually carrying out most aspects of biosecurity recommended by government experts.

BELOW The best way to check the health of your birds is simply to sit and watch them for a time each day.

HANDLING CHICKENS

You will need to handle your chickens from time to time, to move them or to inspect them at closer quarters. First, however, you need to catch them. You may be able to do this by putting down some food and then quickly grabbing them by the legs, or you may be able to drive one bird into a corner and catch it that way. If all else fails, you will be able to catch them once they have gone in to roost for the night.

Birds should never be handled roughly as this will cause stress and may even damage them physically. To pick up a bird, place both hands over its wings, then lift and place it under your arm so that it is facing backwards. Keep hold of both legs with your hand and keep the wings firmly closed between your elbow and your body.

MINOR AILMENTS AND CONDITIONS

If you notice anything a little out of the ordinary, it is important not to panic. Your first approach should be to seek the advice of an experienced poultry keeper, preferably by getting them to visit, although a simple phone call may provide the reassurance you need. If you are still unsure about the diagnosis or treatment of an ailment, do not hesitate to contact your vet.

Colds

It is possible for chickens to catch colds, although the classic symptom of a runny nose may also be an indicator of other problems. As with humans, the likelihood of infection is increased by exposure to draughts, damp and sudden fluctuations in temperature. Colds will get better on their own, but it is possible to purchase avian-cold cures from your vet or agricultural supplier.

Infectious coryza is a bacterial condition that produces symptoms similar to the common cold, including laboured breathing and swollen eyelids. Although this ailment is infectious, an antibiotic treatment is very effective.

Crop-Bound Birds

The crop of a chicken can become congested, or crop bound, for a variety of reasons, the most common being an obstruction caused by eating feathers, litter or long grass. Old-fashioned cures include giving the hen a drink of warm water to distend the crop, which should then be softened by rotating and massaging it by hand. If this does not work, a bird can sometimes be operated on, although it is often necessary to destroy it and open up its crop for analysis. In this way, the problem can be identified and steps can be taken to prevent it from happening to any other birds in the flock.

Egg-Eating

See box on page 152.

Feather-Pecking

Typically, feather-pecking is the result of overcrowding and boredom. By giving your birds more space, dusting sites and a regular supply of greenstuffs, the habit can usually be eradicated, although once it takes hold it is sometimes difficult to stop. Very often, just one bird starts feather-pecking as a result of an inquisitive peck, and if blood is drawn she and others in the flock will be encouraged to continue. At the very least, the neck, rump and vent will be denuded of feathers, and at worst, the flock will continue pecking until the victim is so severely injured that she dies. Pecked birds can be sprayed with proprietary brands of anti-pecking remedies and the wounds treated with an aureomycin-based animal powder.

Fleas

Hen fleas are visible to the naked eye. Flea infestation in a bird is sometimes evident from weight loss and the appearance of bare patches in the feathers. The comb and wattles may also appear anaemic owing to blood loss caused by the feasting fleas. Fleas thrive in warm, humid conditions and are therefore more of a summer problem. Treat birds with a flea powder and spray all internal surfaces of the house to eradicate the flea eggs, pupae and larvae. A regular reapplication is recommended – as often as every three weeks in hot weather.

Lice

These parasites come in several different varieties, each of which has a preference for a certain area of the bird, but they are most commonly found around the vent area (and in the tufts of crested breeds). By parting the feathers, you may see small, light-coloured lice running between them. Look out for lice eggs in the form of a greyish-white encrustation around the vent and in the feathers under the wings.

Treat affected birds with either a liberal dousing of louse powder or a liquid chemical spray. A weekly dusting of nest-boxes will protect laying hens from becoming infested, but any cockerels will have to be treated individually.

Mites

There are several species of mite, the most common being the red mite. Despite its name, this parasite is, in fact, greyish-white in colour, becoming red only when it is full of blood after feeding on birds during the night. Red mites live and breed in crevices in the hen-house as near to their meal as possible, hence their fondness for the ends of perches and the interior of nest-boxes. Watch out for the characteristic signs of infestation – 'egg spotting', or blood spots from the squashed mites on the surface of eggs – and inspect the sleeping quarters regularly by torchlight.

Northern mites are similar in size to red mites but are grey or black in colour and live continuously on the bird's body, most often around the vent area. Birds suffering from a really heavy infestation also tend to become scabby around the facial parts. The red mite and northern mite can both be treated with a pyrethrum-based spray, although a single treatment is seldom effective and regular follow-ups are essential.

The scaly-leg mite differs from other mite species in that it affects only the legs and feet. Left untreated, the legs may become swollen and crusty, and eventually the bird will become lame. The mite burrows under the scales on the legs, creating tunnels where it is able to breed. As it does so, the scales lift and distort. Old-fashioned remedies include sulphur ointment, paraffin and linseed oil. Petroleum jelly also works well, as it seals the gaps in the scales and prevents the mite from breathing. Vets recommend an application of gamma benzene hexachloride.

Prolapse

This condition is rarely seen in intensively-reared birds but is quite common in otherwise healthy free-range birds. It occurs in immature or slightly overweight hens that have been laying heavily, and is caused by straining, the result being that the vent muscles are pushed out of the body with the egg still attached inside.

If prolapse occurs, the egg should gently be broken and removed, and the exposed organs then cleaned with a mild antiseptic. They can then be coated with a lubricant and pushed carefully back inside the abdominal cavity. Keep the hen isolated in a small coop and run for a week, feeding her a diet of grain-based feed. Provided that the problem does not reoccur, the hen can then be returned to the flock.

Worms

Chickens are affected by two main types of worm: the tapeworm, which is flat and segmented; and the roundworm, which is round and smooth. The life cycle of each parasite usually involves a stage in the environment or in an intermediate host. However, most are passed from one bird to another by means of its droppings. Tapeworm eggs may also pass via the droppings or be retained within the rear segments of the worm, which periodically break off and are excreted. These eggs are then eaten by snails and beetles, which in turn are eaten by the poultry.

Birds suffering from worms may show an increase in appetite combined with a decrease in egg production. Combs will look pink rather than red, and birds with a heavy infestation of roundworms will excrete bundles of dead worms. Preventative worming is normally carried out twice a year by the inclusion of a drug in the feed. Flubendazole is the most commonly used vermicide and can be prescribed by a veterinary surgeon, pharmacist or specialist agricultural merchant.

MORE SERIOUS ILLNESSES

Any chickens showing obvious signs of serious ill health must be taken to the vet immediately, and any that die should undergo a post-mortem by a vet who specializes in poultry to determine the cause of death. Unless you are absolutely certain that you have made the correct diagnosis, do not attempt any treatment. While antibiotics are invaluable when used correctly, they should never be given as a standard treatment, as they may merely mask the true symptoms of a disease rather than cure it. In the UK two diseases affecting chickens – avian influenza (bird flu) and Newcastle disease – are notifiable, which means that if you suspect an outbreak of either of these you must, by law, inform the Department for Environment, Food and Rural Affairs (Defra). It is advisable to keep abreast of any developments concerning these diseases, either through the media or via the statements periodically issued by interested organizations and the government's chief veterinary officers.

Acute Death Syndrome

The term 'flip-over syndrome' accurately describes this condition of acute heart failure. The causes of heart attack can be many and varied, and as a result they often remain unknown, but overfeeding is undoubtedly a contributing factor – the syndrome is not uncommon in commercially produced table birds prior to slaughter. If, in post-mortem examination, a fatty deposit is noticed around the heart, liver and kidneys, you have a legitimate reason to suspect overfeeding and should alter the diet of the rest of the flock accordingly.

Aspergillosis

This condition results from the inhalation of large numbers of fungal spores, which create lesions in the respiratory tract and cause obvious respiratory distress. Occasionally, an infection of the brain is also noticed in post-mortem examination. There is no treatment, so it is important to clean out damp feed, long cut grass and mouldy hay, straw or wood shavings from the run, all of which form the perfect mediums for growth of the various associated fungi.

Avian Influenza (Bird Flu)

Bird flu has been given a separate section here owing to worldwide concern over the disease at the time of writing. The disease has actually been recognized since 1878 and by 1901 its cause had been identified as a virus and it was named 'fowl plague'. In 1955, its relationship to the mammalian influenza A viruses had been proven, but it was not until the 1970s that it was realized that vast pools of influenza A viruses also exist in the feral bird population. It is this latter fact that makes the control of any serious outbreak more difficult, especially where poultry is kept outdoors.

The avian influenza virus, known as H5N1, breeds in the respiratory and intestinal tracts of infected birds and is transmitted from bird to bird, either by coughing and sneezing or through the faeces. There is currently little evidence of an airborne spread over long distances. The clinical signs are a cessation of egg-laying, a difficulty in breathing or, sudden death.

Routine biosecurity precautions at home to prevent the spread of bird flu include keeping visitors away from your birds, ensuring that all feeding utensils in the hen-house are kept clean in order to reduce visits by wild birds, and the addition of vanodine or a similar product to the drinking water each day. Spraying the insides of the house on a weekly basis with a virucide will also help.

A vaccine containing an inactivated strain of the H5N2 strain is available against bird flu. The antibodies produced by birds given this vaccine are effective against H5N1 because the two strains are so similar. Although birds do need time to build up immunity following vaccination, tests have proven that transmission of the virus is halted after two weeks. The duration of protection is further increased by a booster vaccination given some six to ten weeks after the initial dose. Some countries have opted for a proactive policy of vaccination of domestic poultry, as this will help to avoid millions of birds having to be destroyed should a serious outbreak of bird flu ever occur. If you have any suspicions that your birds might be showing signs of bird flu, you must notify your vet or the appropriate government authority immediately.

Coccidiosis

Normally affecting only young stock, coccidiosis is bleeding within the intestinal walls and is caused by the coccidial protozoan organism. It is spread though droppings, so a good cleaning regime should prevent an outbreak. A natural immunity also tends to develop within a flock, and it has been found that young birds raised on wire floors, where they have no contact with their droppings, are more susceptible. Symptoms include a hunched appearance, ruffled feathers, blood in the droppings and/or sticky white diarrhoea coating the feathers of the vent area. Adult birds may lose weight and condition, bringing about a decrease in egg production.

Chick crumbs and growers' pellets contain anticoccidiostats that normally protect against the disease, although a vaccine is also available. Alternatively, it is possible to introduce coccidials to young birds at an early age so that they can build up a lifetime's immunity. If a problem occurs in untreated adult birds, your vet should be able to provide you with a prescribed drug.

Coronavirus

Several types of coronavirus cause disease in poultry, including infectious bronchitis, which results in respiratory problems and kidney damage. Generally, the virus causing infectious bronchitis is airborne, spreads rapidly throughout a flock and can persist in an individual bird for several months. The first signs of infection are a drop in egg production and the appearance of rough, wrinkled shells on eggs that are laid. It is possible to vaccinate against the various coronavirus diseases, and good biosecurity will help prevent their spread to neighbouring flocks.

Marek's Disease

There are several forms of this viral disease, a kind of herpes. The classic form appears as lameness in one leg, with the wing dragging on the ground. Some chickens will limp for a time then appear to recover completely, but this is, unfortunately, rare. Some breeds – such as Sebrights and Barnevelders – appear to be more susceptible, whereas Marans and Sumatras are hardly ever affected.

Stress is often the trigger, and pullets are particularly vulnerable at point-of-lay. A high incidence of parasitic worms and introducing the bird to a new environment can result in onset of the disease, which may hitherto have been dormant.

It is possible to vaccinate against Marek's disease, but obtaining small quantities of vaccine may be a problem for the average poultry keeper. An answer is to buy birds that have already been vaccinated, or those that have been reared under a broody hen, as these seem to succumb to the disease only very rarely. Alternatively, consider clubbing together with other poultry keepers in a collective vaccination programme.

Newcastle Disease (NCD, or Fowl Pest)

Along with avian influenza, Newcastle disease is classed as 'notifiable' owing to its extremely contagious nature and potentially high mortality rates, although outbreaks can vary from very mild to severe. The disease produces various symptoms, including respiratory and nervous disorders, diarrhoea, lethargy and depression. In young birds, Newcastle disease can cause a severe reduction in growth and a number of secondary diseases.

It is spread by direct contact with secretions and excretions, especially faeces, and also via contaminated feed and water vessels. Because it is so contagious, in some countries it is compulsory that birds are vaccinated before being allowed to enter shows. As with Marek's disease, it might pay to form a combined vaccination group with other poultry keepers to reduce costs.

SHOWING

FOR SOME POULTRY KEEPERS, EXHIBITING BIRDS IS THE ONLY REASON FOR CHOOSING A PARTICULAR BREED. SHOWING IS NOT ONLY FUN, BUT ALSO AN EXCELLENT WAY OF KEEPING IN REGULAR CONTACT WITH OTHER CHICKEN FANCIERS AND BREEDERS OF GOOD-QUALITY BIRDS. INCIDENTALLY, THERE IS NOTHING WRONG WITH SHOWING BIRDS THAT YOU HAVE NOT BRED YOURSELF – PROVIDING THAT THEY HAVE BEEN IN YOUR POSSESSION LONG ENOUGH TO QUALIFY UNDER THE RULES AND REGULATIONS SET OUT IN THE SHOW SCHEDULE – BUT YOU WILL UNDOUBTEDLY GAIN A GREAT DEAL MORE SATISFACTION IN WINNING WITH YOUR 'OWN' BIRDS.

BIRD CARE

If you keep birds for exhibition, they will require some extra care in the days before a show to ensure that they are looking their best.

WASHING BIRDS

Only birds that are to be exhibited are normally washed, and of these only the soft-feathered breeds. Washing hard-feathered varieties such as the Old English Game spoils the tightness of the feather conformation, so traditionalists 'polish' them with a silk handkerchief instead. Washing can also occasionally be necessary in breeds that have crests or feathered feet.

For showing purposes, washing should be carried out several days before the event so that the feathers have time to dry fully and hang properly. Most fanciers use two bowls of lukewarm water, the first ideally containing pure soapflakes, although washing-up liquid or baby or dog shampoo will do almost as well. Whatever soap is chosen, use only a little or it will strip too much of the natural oils from the feathers. The second bowl should contain clean water for rinsing, although the addition of some clothes whitener can be beneficial with white or light-coloured breeds.

First, wash the feet and legs separately outside the main bowl, so that the dirt does not go into the clean water. A firm toothbrush or nailbrush is ideal for this, but any dirt that is ingrained under the scales can be picked out using a cocktail stick or toothpick.

Next, holding the bird firmly so that it feels secure, immerse it in the first bowl of water and, starting at the head, very gently work the lather through the feathers in their direction of growth, taking care not to get soap in the bird's eyes. Once they have got over the initial panic of being placed in the water, most chickens seem to like this part and will crouch quietly in the bowl while they are being washed.

When the bird has been rinsed clean in the second bowl of water, pat it dry with kitchen roll or old towels. Some exhibitors use a hairdryer, although if held too closely this can damage the feathers. The bird can then be left to dry completely in a box or poultry basket in a warm room or near a radiator. Once it is thoroughly dry, a show bird should then be put in a show pen in the penning room. If it is not intended for exhibition, the bird can be returned to the flock, although remember that the washing process will have removed some of the feathers' natural oils and so the bird may have less protection against inclement weather for a day or two.

SHOWING TIPS

Every experienced poultry exhibitor has their own methods of enhancing a bird's appearance. For some this involves shaping and drying the feathers in a particular way after washing, while others may use rouge or baby oil to brighten a red comb, baby powder to whiten ear lobes or petroleum jelly to smarten up legs. There are a multitude of other tips that you will find out by befriending a successful breeder, although they are unlikely to divulge any ingenious tricks of the trade if you are going to be in direct competition.

SHOWING

It is best to start with small local shows, which are organized by poultry societies or bantam clubs. As you gain experience (and provided that your birds are good enough), it may eventually be worth considering regional or annual shows, where there may be several thousand birds competing under one roof. The main period for showing is winter, because chicks that hatch in early spring are fully grown in mid-autumn and by then yearling birds will have recently acquired their new plumage.

Entry forms and schedules are normally available a couple of months before the show date. The schedule will contain all you need to know: date and location, the cost of entry, the types of classes and, most importantly, the closing date for entries. It should also tell you when you can pen your birds and the earliest hour that you will be allowed to take them home. It is important that you read the schedule carefully and enter your birds in the correct classes. Birds that are going to be exhibited need to be chosen with care and prepared thoroughly before the event. They must conform to breed standards and they also need to be 'trained' to show themselves to their best advantage.

BREEDING A WINNER

Breeding a show winner can take years and, despite buying the best stock, seeking the advice of more experienced handlers and training their birds to stand well in an exhibition pen, many have tried and failed to take home that elusive red card and ultimate trophy of 'Best in Show'. Of course, buying the best breeding stock, taking advantage of

RIGHT Start small, with local shows, and consider joining a breed club or society.

experienced handlers and preparing properly helps, but it offers no guarantees. To produce a show winner, it is first of all necessary to understand the standards required for each breed. In the UK, these can be found in the book of British Poultry Standards, which is normally published every ten years. In the US they are found in The Standard of Perfection, produced by the American Poultry Association.

While a bird's overall breed characteristics are an obvious requirement, the newcomer should also consider the very real need for correct shape, colour and markings. If you don't understand exactly what is required, you could end up getting rid of potential show winners and keeping (and subsequently attempting to show) the less than perfect offspring that result from an initial mating. Before even considering breeding for showing, join your local poultry club and any national clubs such as the Poultry Club of Great Britain or the American Poultry Association and enlist the support and constructive criticism of an expert.

If your birds have been taught to stand well in the show cage, they are far more likely to catch the eye of the judge than if they are huddled up in one corner. Unlike dogs or other livestock which can be walked and trotted around a show ring in order to show their conformation, chickens can be assessed only picking up each bird and handling it. So yes, the judge will take a closer look later and will evaluate all the characteristics required by your particular breed, but if it comes to the point where two birds have amassed an identical number of marks, it will be the one that has that

LEFT As with any sort of showing, whether it is produce or livestock, first impressions count.

'certain something' about its stance and bearing that will be placed first.

It therefore pays to 'train' your birds by placing them in pens of similar dimensions to show cages for a few days before the event. You can buy show-pen fronts and make your own – it doesn't have to be a beautifully crafted piece of engineering; the point is merely to accustom your bird to it so that it feels at home and happy. If you have a suitable well-ventilated food store, find an old table and place your practice pen on that. In a very short time, the occupant will become used to you going in and out and brushing past in close proximity – all experiences which will help it to be confident on the day. Judges very often carry a small telescopic stick (much like a car aerial) which they poke gently through the bars in order to get a show bird to stand properly. You might like to keep a thin garden cane by the side of your makeshift show pen and occasionally do the same – it all helps and such simple things can make all the difference between being in the top three or down there with the 'also-rans'. It is of course important that your pens have food and water vessels attached to the bars to avoid mess: it is likely that by the time the birds are in your practice pen they will have been washed, primped and preened, and you do not want their plumage becoming soiled.

ON THE DAY

If the venue isn't nearby, be sure to allow plenty of travelling time and plan your route carefully in advance. Before leaving home, double check the latest times at which your birds must be in their pens and give yourself an extra half-hour or so to allow for traffic hold-ups and punctures.

On arrival at your first show, don't panic! See the stewards, obtain your pen number and, before you put the bird into the cage, take time to complete those last-minute preparations advised by those more experienced enthusiasts at your local club.

Finally, if you are not successful, lose with grace. You may be lucky enough to find your bird has won a rosette and place card but if not, pick up another schedule and enter somewhere else in a few weeks' time. If you see the judge somewhere in his or her white coat, ask if they have the time to come to where your bird is penned and explain the 'faults' that prevented it from being placed. Never be a bad loser, and never forget that although showing is great fun and can be addictive, it is only a part of the pleasure to be gained from keeping poultry.

LEG RINGS

You may notice some chickens in the show pens that are sporting leg rings. While some countries have operated a voluntary ringing scheme (usually organized by the national poultry club) for some time, the question of compulsory leg rings for show birds divides enthusiasts in countries where it is not already a 'done deal'. In the UK, opponents of the scheme describe it as 'bureaucratic and unnecessary', whereas those in its favour believe that it is an excellent way of recording the pedigree and ownership of an individual bird. The basic idea behind the scheme is that by registering birds with a central body, it should always be possible to prove the breed and the bird as well as going some way towards offering a degree of traceability should a bird be stolen (although there would be nothing to prevent a thief from removing the ring, the fact that

it had no ring would in itself be suspicious). At the moment, anyone can, quite easily, enter their birds into a show but the fear is that if eventually an edict makes leg rings compulsory then the new fancier would find it more difficult to begin a show career. There might also be the problems currently being experienced by European pigeon exhibitors whereby birds ringed in one country cannot be shown in another.

AFTER THE SHOW

Once you get your birds home, they should always be quarantined for a few days before being returned to the flock to prevent the possible transmission of disease.

Anyone who is interested in showing chickens should read David Scrivener's comprehensive and informative book *Exhibition Poultry Keeping* (Crowood Press, 2005).

GLOSSARY

Addled A fertile egg, the embryo of which has died during incubation.

Air sac The air space found at the broad end of the egg. It denotes freshness and, during incubation, the development of the embryo.

Alektorophobia The fear of chickens.

Auto-sexing Used of breeds in which the sex of day-old chicks can automatically be determined by the markings on their down. See also 'Sex linkage'.

Axial feather When the wing of a chicken is opened out, you will see along the edge, a single layer of feathers sticking out. Roughly in the middle is a shorter feather, and this is the axial feather, showing where the primary and secondary feathers begin and end.

Beard See 'Muff'.

Blood spot Seen in the laid egg; its appearance could be due to generic, nutritional causes or the hen experiencing sudden stress.

Bloom The moist, protective layer on a newly laid egg. Also, a description of the feathers of a bird in prime and/or exhibition condition.

Booted Having feathers on the feet and legs.

Boule Type of feather formation found on the necks of some continental breeds.

Brassiness Discolouring in light-coloured breeds, caused by sunlight and weathering.

Broiler A young, tender chicken specifically intended for the table.

Brooder An artificial heater used for rearing young chicks.

Broody As an adjective, having the natural instinct to sit on eggs once a clutch has been laid. As a noun, a hen bird showing the same inclination, which will sit either on her own eggs or on those substituted.

Caecum One of two intestinal pouches found at the junction of the small and large intestines.

Candling Using a light to examine the contents of an egg without breaking the shell, often to see if embryos are still alive in the egg.

Cape Feathers running from the back of the head down to the shoulders.

Capon A castrated male chicken.

Carrier Either an outwardly healthy chicken that can, nevertheless, pass on disease to other poultry in the flock; or, more obviously, a box in which chickens are transported.

Clutch A batch of eggs that are incubated together; or, all the eggs laid by a hen on consecutive days before she skips a day and starts a new laying cycle.

Coccidiosis A parasitic protozoal infestation.

Coccidiostat A preventative drug used to keep chickens from contracting coccidiosis.

Cock Mature male after its first breeding year.

Cockerel Male bird before it becomes a cock. Also used as a generic term for male chickens of any age, especially in the UK.

Conformation The shape of a chicken; the standards for which of all pure breeds are laid out in books normally published by the poultry clubs of each country.

Crest Tuft of feathers on the heads of some breeds. Sometimes known as a 'tassel'.

Debeaking Trimming back a bird's upper beak to prevent it from feather-pecking.

Droppings board A removeable board fitted under the perches to collect faeces.

Dubbing The removal of the male bird's comb and wattles, first carried out during the days of cock fighting to prevent injury. Until recently, show strains of some breeds – such as the Old English Game – were still dubbed, but the practice is now either illegal or frowned upon in most countries.

Duck-footed A serious show fault where the rear toes are out of line.

Ear lobes The patches of skin below a chicken's real ears. In most cases, the colour of ear lobes indicates eggshell colour: white-lobed birds lay white or cream eggs, while red-lobed birds lay brown or tinted eggs.

Ear tufts Clumps of feathers growing from small tabs of skin usually found at or near the region of the ear openings.

Egg tooth Found on the upper part of a chick that has just hatched and used to chip through the eggshell. It is lost after a few hours.

Embryo The developing chick inside an egg.

Flight feathers The large primary feathers on the last half of the wing.

Fold unit Portable combined house and run.

Force moult Artificially induced to moult at an unnatural point of a bird's cycle. Sometimes carried out to ensure a bird is in prime condition for a show on a given date.

Free-range Access to the outdoors, sometimes restricted, sometimes unrestricted.

Frizzled Where each feather curls up so that its tip points towards the bird's head.

Fryer See 'Broiler'.

Gizzard A grinding stomach with a muscular lining.

H5N1: A particular strain of avian influenza (bird flu).

Hackles See 'Cape'.

Heavy breed A breed whose ancestry possibly derives from Chinese fowl.

Hen Female after her first laying season.

Immunity Resistance to a particular disease.

Impaction Blockage of a particular part of the body, normally used in connection with problems appertaining to the crop.

Juvenile The first set of proper feathering to moult out before the adult feathers appear. In showing, a juvenile category refers to the age of the exhibitor, not the age of the chicken.

Keel The bony ridge of the breastbone.

Light breed A breed that is usually Mediterranean in origin or has jungle fowl ancestry.

Litter Covering for the floor and/or nest boxes. Normally wood shavings or straw, but can be dried leaves, shredded paper, flax or any clean absorbent material.

Mille fleur A colour pattern, not a breed.

Mite Tiny parasitic insect that lives upon the skin and feeds from the body.

Meat spot Small deposit of blood in an egg, caused by the rupture of a blood vessel during formation.

Moult The period when a chicken sheds its old feathers and grows new ones.

Muff Puff of feathers on either side of the face. Usually seen in conjunction with a beard.

Neck moult Moulting of the neck feathers only; it often occurs when a pullet starts laying before she is fully mature.

Oil sac A gland at the base of the tail that contains oils used in preening to keep the feathers in good condition.

Oviduct The tube down which an egg travels when it is ready to be laid.

Parasite Any insect that feeds on the body of a host, taking nutrition and shelter from the host or its immediate surroundings such as the end of perches and dark corners of nest boxes.

Pigmentation The colour of a chicken's beak, legs and vent.

Parson's nose The protruding lump of flesh normally covered by the tail feathers and seen only on a plucked carcass.

Point-of-lay (POL) The time at which a hen bird is expected to commence laying, usually anywhere between the ages of 18 and 22 weeks, although some breeds will start laying later than this.

Primary feathers The long, stiff feathers at the outer tip of the wing.

Pullet Young hen from hatching to the end of the first laying season.

Rust A show fault, denoting an undesirable reddish area on the wings of partridge, duckwing or pile-coloured females.

Saddle Part of the chicken's back just before the tail.

Scales The covering on the legs and feet.

Scratch Grain fed to chickens, normally as an afternoon feed.

Secondary feathers The feathers on the outer side of the wing situated between the primaries and the point at which the wing joins the body.

Setting The act of putting fertile eggs under a broody hen or in an incubator to hatch them.

Sex linkage When the sex of chicks can be distinguished at the time of hatching by their appearance. If, for example, a gold male is mated to silver females, the pullet chicks will follow the colour of the father while the cockerel chicks will be almost wholly silver.

Shank The part of the leg between the claw and the first joint.

Sickles Long, curved feathers on the outer side of a cock bird's tail.

Spur Pointed, horny projection at the base and rear of a cock bird's legs. Contrary to popular opinion, the length of the spurs is not a reliable indicator of a bird's age.

Standard Either conforming to the standard i.e. perfect specimen of the breed as laid down by the various poultry clubs; or 'standard-sized' – large fowl as opposed to bantams.

Starve-out Failure of newly hatched chicks to eat after absorbing the egg sac.

Sternum See 'Keel'.

Strain A group or flock of chickens carefully bred over several generations by an individual fancier.

Topknot See 'Crest'.

Trachea The windpipe, which forms part of the respiratory system.

Type The typical shape and colour of a chicken indicating the actual breed.

Undercolour The colour of the fluff and lower parts of the feathers.

Variety A particular type of a pure breed – bantam Partridge and large Whites are, for example, both varieties of the Wyandotte breed.

Vent The orifice at the rear of a bird, also called the cloaca, through which droppings and eggs are passed.

Wattles The fleshy parts immediately below the head.

Wing clipping Clipping the primary and secondary feathers of one wing to unbalance a bird and so prevent it from flying.

PICTURE CREDITS

USEFUL ADDRESSES

Poultry Club of Great Britain
Keeper's Cottage
40 Benvarden Road
Dervock
Ballymoney
Co. Antrim
BT53 6NN
www.poultryclub.org

Rare Breeds Survival Trust
National Agricultural Centre
Stoneleigh Park
Warwickshire
CV8 2LZ
www.rbst.org.uk

Rare Poultry Society
www.rarepoultrysociety.co.uk

Utility Poultry Breeders
Association
Morville Heath
Bridgenorth
Shropshire
www.utilitypoultry.co.uk

American Bantam Association
PO Box 127
Augusta, NJ 07822
www.bantamclub.com

American Poultry Association
PO Box 306
Burgettstown, PA 15021
www.amerpoultryassn.com

INDEX